Contents

Welcome & What You'll Learn

Welcome, intrepid navigator, to the exciting world of Agile! Whether you're a seasoned project veteran or a wide-eyed beginner, this book is your map to harnessing the power of Agile methodologies. Here, you'll shed the rigid confines of traditional project management and dive into a dynamic, collaborative, and **rapidly evolving landscape** where flexibility reigns supreme.

What Is Agile, Exactly?

Imagine a project that adapts to change like a chameleon, thrives on continuous feedback, and delivers value in bite-sized increments. That's the essence of Agile. Forget about bulky project plans and waterfall stages; Agile embraces **iterative cycles**, delivering working software early and often, allowing you to learn and adjust course with each step.

Why Should You Care?

In today's fast-paced world, the **traditional "plan-then-execute" approach** often stumbles. Requirements shift, markets evolve, and deadlines loom, leaving even the best-laid plans in disarray. Agile empowers you to **embrace this uncertainty**, turning it into a fuel for innovation and resilience.

What Will You Learn Here?

This book is your comprehensive guide to **setting up, executing, and excelling** in Agile projects. We'll delve into:

- **Agile Demystified:** Unpacking the core principles and values that drive this dynamic approach.
- **Traditional vs. Agile:** Contrasting the two methodologies, highlighting Agile's advantages in adaptability and responsiveness.

- **Picking Your Agile Framework:** Exploring popular frameworks like Scrum, Kanban, and Lean, equipping you to choose the best fit for your project.
- **Building Effective Agile Teams:** Assembling cohesive, high-performing teams that thrive on collaboration and shared ownership.
- **Nurturing Agile Teams:** Mastering communication, conflict resolution, and the art of giving and receiving productive feedback.
- **Estimation Techniques:** Demystifying agile estimation, employing practical techniques to accurately gauge workload and set realistic expectations.
- **Planning for Flexibility:** Crafting adaptable plans that embrace change and empower your team to make informed decisions.
- **Executing Agile Iterations:** Understanding the structure and rhythm of sprints, the building blocks of Agile execution.
- **Conquering Iteration Challenges:** Addressing common roadblocks like scope creep, communication gaps, and motivation dips.
- **Mastering Agile Communication:** Refining your communication skills for effective daily standups, retrospectives, and feedback loops.
- **Agile in Software Development:** Deep-diving into how Agile principles empower software development through practices like unit testing, refactoring, and continuous integration.

Beyond the Book:

This book is just the beginning! To further enrich your Agile journey, consider exploring these resources:

- **Agile Alliance:** https://www.agilealliance.org/ - A treasure trove of articles, guides, and events.

- **Scrum Guides:** https://scrumguides.org/ - Official documentation for the popular Scrum framework.
- **Kanban University:** https://kanban.university/ - Dedicated to all things Kanban, with insightful articles and training.
- **Agile Podcasts:** Podcasts like "Agile for Humans" and "Agile Revolution" offer engaging insights and interviews with Agile experts.

Get Ready to Embrace the Agile Revolution!

By the end of this book, you'll be equipped to navigate the world of Agile with confidence, leading your team to deliver value, foster collaboration, and conquer project challenges with a **smile on your face and a spring in your step**. So, buckle up, open your mind, and get ready to embrace the Agile revolution!

Section 1:
Understanding Agile

Agile Demystified: An Overview

Welcome to the world of Agile! For some, it's a revolution in project management, a breath of fresh air in our process-driven world. For others, it's a confusing jumble of jargon and buzzwords. This chapter aims to bridge this gap, stripping away the mystery and presenting Agile in its clear, practical essence.

1.1 Beyond the Buzzwords: What is Agile?

Imagine a chef cooking a delicious meal. Instead of following a rigid recipe, they adjust on the fly, tasting and adapting as they go. This is the core of Agile: **an iterative and incremental approach to project management**. We break down our goals into small, manageable chunks, working in **short, focused cycles called sprints**. After each sprint, we gather feedback, learn, and adapt, constantly refining our work towards the final outcome.

1.2 Embracing Flexibility: Key Pillars of Agile

Four core values guide the Agile philosophy:

1. Individuals and interactions over processes and tools:
People are at the heart of Agile. Collaboration, open communication, and trust are paramount, not rigid methodologies or fancy software.

2. Working software over comprehensive documentation: We prioritize delivering functional pieces of the project over extensive documentation. Think of building a bridge one section at a time, ensuring each section is strong before moving on.

3. Customer collaboration over contract negotiation: Continuous user feedback is crucial. Instead of locked-in contracts, we embrace feedback loops, adapting to changing needs and delivering maximum value.

4. Responding to change over following a plan: Flexibility is key. While initial plans provide direction, we readily adapt to new information and changing priorities, ensuring our project stays relevant and impactful.

1.3 Agile Frameworks: Navigating the Options

Agile isn't one-size-fits-all. Several frameworks tailor these principles to specific contexts. Here are some popular options:

- **Scrum:** Simple and structured, with defined roles (Product Owner, Scrum Master, Development Team) and time-boxed sprints.
- **Kanban:** Visualize workflow with boards and cards, focusing on continuous flow and optimizing work-in-progress.
- **Lean:** Minimize waste and maximize value through iterative cycles and continuous improvement.

Choosing the right framework depends on your project's complexity, team size, and desired level of structure.

1.4 Debunking the Myths: Common Misconceptions about Agile

Myth #1: Agile is chaotic and lacks planning.

Reality: Agile embraces flexibility, not anarchy. Planning still exists, but it's adaptable and ongoing, adjusting to new information and feedback.

Myth #2: Agile means no documentation.

Reality: While prioritizing working software, Agile recognizes the value of documentation. Documentation is kept lean and up-to-date, reflecting the current state of the project.

Myth #3: Agile is only for software development.

Reality: Agile principles can be applied to any project, from marketing campaigns to product launches to even running a household!

1.5 Taking the First Step: Embracing the Agile Mindset

Remember, Agile is a journey, not a destination. It's a shift in mindset, embracing collaboration, continuous learning, and a willingness to adapt.

Here are some tips to get started:

- **Read and learn:** Dive into Agile resources like the Agile Manifesto (https://agilemanifesto.org/) and explore different frameworks.
- **Start small:** Implement Agile principles in small projects to gain experience and build confidence.
- **Find your community:** Connect with other Agile practitioners, share experiences, and learn from each other.

So, are you ready to embrace the flexibility and adaptability of Agile? Dive into the next chapters to unlock the power of this transformative approach and lead your project to success!

Additional Resources:

- **Agile Alliance:**
  https://www.agilealliance.org/agile101/the-agile-manifesto/
- **Scrum Guides:** https://scrumguides.org/
- **Kanban University:** https://kanban.university/
- **Lean Enterprise Institute:** https://www.lean.org/
- **Atlassian Agile Coach:**
  https://www.atlassian.com/agile/about

Agile vs. Traditional Project Management: Key Differences

Welcome to the world of Agile! Before diving headfirst into its methodologies and intricacies, let's equip ourselves with some crucial knowledge – understanding how Agile differs from traditional project management approaches. This comparative understanding will not only solidify your grasp of Agile but also help you decide if it's the right fit for your project.

Traditional Project Management

Imagine a meticulously crafted blueprint, complete with detailed specifications and a rigid timeline. That's the essence of traditional project management. Think Waterfall, the most common method, where projects progress sequentially through distinct phases – planning, design, development, testing, and deployment.

Key features of traditional project management:

- **Upfront planning:** Extensive planning at the beginning defines the entire project scope, timeline, and budget, leaving little room for adjustments.
- **Silos and hierarchies:** Teams are often structured in departments, with clear roles and reporting lines.
- **Focus on deliverables:** The emphasis lies on meeting pre-defined milestones and deadlines, sometimes at the expense of flexibility or user feedback.
- **Limited iteration:** Changes are challenging and often require restarting phases, leading to slower adaptation.
- **Documentation-heavy:** Detailed documentation for requirements, specifications, and designs is crucial.

Agile Project Management

Now, picture a team of skilled climbers navigating a dynamic mountain face, adapting their route based on real-time feedback and changing conditions. That's the spirit of Agile! It's an iterative and incremental approach, valuing flexibility, collaboration, and continuous improvement over rigid plans.

Key features of Agile project management:

- **Iterative development:** Work is broken down into short cycles (sprints) with regular delivery of working software increments.
- **Cross-functional teams:** Teams are self-organizing and collaborative, with members from different disciplines working together.
- **Focus on value:** The priority is on delivering value to users early and often, incorporating feedback to continuously improve the product.
- **Embrace change:** Adaptability is key, with projects evolving based on feedback and new learnings.
- **Minimal documentation:** Focus on just-enough documentation that supports collaboration and knowledge sharing.

So, what are the key differences?

Let's compare these two approaches side-by-side:

Feature	Traditional Project Management	Agile Project Management
Planning	Upfront and detailed	Iterative and incremental
Team structure	Silos and hierarchies	Cross-functional and self-organizing

Focus	Deliverables and milestones	Value and user feedback
Change	Difficult and costly	Embraced and encouraged
Documentation	Extensive and comprehensive	Minimal and just-enough

Choosing the right approach

The best approach depends on your project's specific needs and complexities.

- **Traditional:** Choose traditional methods for well-defined projects with stable requirements, predictable environments, and a need for upfront control.
- **Agile:** Embrace Agile for projects with evolving requirements, high uncertainty, rapid feedback loops, and a focus on continuous value delivery.

Remember, Agile isn't a one-size-fits-all solution. Different Agile methodologies like Scrum, Kanban, and Lean offer variations, allowing you to tailor the approach to your specific needs.

Additional Resources:

- Atlassian Agile Coach's Handbook: https://www.atlassian.com/agile/about
- The Agile Manifesto: https://www.agilealliance.org/agile101/the-agile-manifesto/
- Scrum Guides: https://www.scrum.org/resources/scrum-guide

I hope this chapter has equipped you with a clear understanding of the key differences between Agile and traditional project management. Remember, choosing the right approach is crucial for project success. So, keep an open mind, explore your options, and embrace the flexibility and adaptability that Agile offers!

Debunking Common Agile Misconceptions

Welcome to the demystification zone! In this chapter, we'll tackle some of the most prevalent misconceptions surrounding Agile. Understanding these false notions is crucial for adopting Agile effectively and reaping its true benefits. So, grab your skepticism hats and let's dive in!

Misconception #1: Agile is all about chaos and no planning.

Reality: While Agile emphasizes flexibility and adaptation, it's not synonymous with chaos. Agile teams prioritize iterative planning, focusing on short sprints and adapting to changing priorities as needed. Think of it as building a ship while sailing, not charting a rigid course years in advance.

Think about it: Imagine planning a road trip across the country. You research the route, book hotels, and pack essentials. But what if you encounter a scenic detour or a sudden downpour? Agile planning allows you to adapt to these unexpected twists while still reaching your destination.

Misconception #2: Agile is just for software development.

Reality: While Agile originated in software development, its principles are remarkably versatile. From marketing campaigns to product launches, Agile can be applied to any project that requires adaptability and continuous improvement.

Think about it: Imagine planning a wedding. You have a budget, a guest list, and a vision. But what if the florist runs out of your favorite flowers or the DJ cancels? Agile principles, like clear communication and iterative planning, can help you navigate these challenges and ensure a beautiful ceremony.

Misconception #3: Agile eliminates the need for documentation.

Reality: While Agile prioritizes face-to-face communication and rapid iterations, documentation remains crucial for knowledge sharing, onboarding new team members, and ensuring project consistency. The key is to find the right balance between detailed documentation and concise, readily accessible information.

Think about it: Imagine building a house. You don't need blueprints for every nail, but you do need a clear plan for the foundation, walls, and roof. Agile documentation should be just as focused, providing a high-level overview and key details without bogging down the team.

Misconception #4: Agile means working longer hours.

Reality: Agile prioritizes short sprints and regular feedback loops, preventing burnout and promoting a healthy work-life balance. The focus is on working smarter, not harder, by eliminating unnecessary tasks and optimizing workflows.

Think about it: Imagine running a marathon. You wouldn't sprint the entire distance, would you? Agile breaks down the project into manageable sprints, allowing for rest and reflection between bursts of activity.

Misconception #5: Agile guarantees success.

Reality: While Agile offers a powerful framework for project management, its success depends on several factors, including team commitment, organizational culture, and project complexity. It's not a magic bullet, but it can significantly improve project outcomes when implemented correctly.

Think about it: Even the best recipe won't guarantee a delicious meal if you don't follow the instructions or use the right ingredients.

Agile is a powerful tool, but its effectiveness relies on the skill and dedication of the team and the suitability of the project.

Remember: Debunking these misconceptions is just the first step. To truly embrace Agile, focus on its core principles: **collaboration, adaptability, continuous improvement, and delivering value.** By building a strong Agile foundation, you'll be well-equipped to navigate the ever-changing project landscape and achieve success.

Additional Resources:

- Agile Manifesto: https://www.agilealliance.org/agile101/the-agile-manifesto/
- Common Agile Misconceptions: https://guidehouse.com/insights/advanced-solutions/2023/debunking-common-agile-myths
- Debunking Agile Myths: https://www.hulhub.com/debunking-common-misconceptions-about-agile

By understanding these resources and debunking common misconceptions, you can confidently embark on your Agile journey and unlock its full potential. Remember, the key is to embrace the flexibility, adaptability, and continuous improvement that Agile offers. So, go forth and conquer your projects, one sprint at a time!

Choosing the Right Agile Methodology for Your Project

Welcome to the crossroads of agility! Having grasped the core principles of Agile, you're now faced with a crucial decision: **selecting the right methodology** for your project. Fear not, explorer, for this chapter equips you with a map to navigate the diverse landscape of Agile frameworks.

Deconstructing the Maze: Popular Agile Methodologies

Before diving into specifics, let's peek at some prominent Agile methodologies that might entice you:

- **Scrum:** The ever-popular champion of iterative development, Scrum thrives on short sprints, self-organizing teams, and clear roles like the Scrum Master and Product Owner. Think of it as a well-oiled machine churning out value in time-boxed sprints.
- **Kanban:** Visualize a flowing river of tasks on a Kanban board! This method prioritizes continuous flow and focuses on managing work in progress (WIP) to avoid bottlenecks. Imagine a team swimming with the current, adapting to changes effortlessly.
- **Lean:** Inspired by Toyota's production system, Lean emphasizes eliminating waste and maximizing value. Think of it as a chisel, meticulously carving away inefficiencies for a streamlined project.
- **XP (Extreme Programming):** Buckle up for a fast-paced journey with XP! This methodology champions continuous feedback, tight collaboration, and rapid testing to ensure quality at every turn. Think of it as a rocket ship, propelling development with constant feedback loops.

These are just a few flavor profiles in the Agile buffet. Other methodologies like Crystal, Feature-Driven Development (FDD), and Dynamic Systems Development Method (DSDM) cater to specific project needs and team dynamics.

Agile Methodology Comparison Chart

Methodology	Project Complexity	Team Size	Stakeholder Involvement
Scrum	Moderate-High	5-9	High (Sprints provide regular visibility)
Kanban	Low-Moderate	3-12	Moderate (Continuous flow updates stakeholders)
Lean	Moderate-High	5-15	Moderate (Focus on value delivery keeps stakeholders engaged)
XP (Extreme Programming)	Moderate-High	3-7	High (Frequent feedback loops)
Crystal	Low-Moderate	5-15	Low-Moderate (Empowers teams for self-organization)

Feature-Driven Development (FDD)	Moderate-High	7-15	Moderate (Feature teams collaborate with stakeholders)
Dynamic Systems Development Method (DSDM)	High	10-20	High (Rapid iterations with stakeholder feedback)

Additional Notes:

- This is a simplified comparison and individual methodologies may have variations in their application.
- Consider the specific needs of your project and team when making your final decision.
- Don't be afraid to experiment and adapt your chosen methodology as needed.

Example:

Imagine you're working on a project to develop a new mobile app. The project has moderate complexity, involves a team of 7 developers, and requires regular input from stakeholders. Based on the table, Scrum or Kanban could be good options:

- **Scrum:** Provides clear visibility through sprints, which might be valuable for stakeholders.
- **Kanban:** Offers flexibility to adapt to changing requirements, which could be beneficial for a new app development project.

Ultimately, the best choice depends on your specific situation and preferences.

I hope this table helps you compare different Agile methodologies and choose the right one for your project!

Finding Your Perfect Match: Choosing the Right Framework

So, how do you pick the perfect Agile partner for your project? Here are some key factors to consider:

- **Project complexity:** Is your project a well-defined beast or a shape-shifting enigma? Complex projects might benefit from the structure of Scrum, while Kanban's flexibility caters well to evolving requirements.
- **Team size and composition:** Are you a lean, mean, coding machine team or a diverse orchestra of skills? Smaller teams might find Kanban's simplicity liberating, while Scrum's roles can provide structure for larger groups.
- **Stakeholder involvement:** Do you have stakeholders eagerly peering over your shoulder, or are they hands-off cheerleaders? Scrum's clear visibility through sprints might appease involved stakeholders, while Lean's focus on value delivery resonates with hands-off supporters.
- **Project timeline and budget:** Are you sprinting towards a tight deadline or meandering on a flexible timeline? Scrum's predictable sprints work well for fixed deadlines, while Kanban's continuous flow adapts to evolving timelines and budgets.

Remember, there's no "one size fits all" solution. Don't be afraid to **mix and match elements** from different methodologies to create a hybrid approach that suits your unique project needs.

A Practical Compass: Tools for Making the Right Choice

Here are some practical tools to guide your decision-making:

- **Agile maturity assessments:** Online quizzes and frameworks can help evaluate your team's Agile readiness and suggest suitable methodologies.
- **Prototyping and experimentation:** Don't be afraid to test-drive different approaches in small pilots. See what resonates with your team and project dynamics.
- **Seeking expert advice:** Consulting Agile coaches or experienced practitioners can provide valuable insights and recommendations.

Decision-Making Process for Choosing an Agile Methodology

Step	Description	Example
1. Define your project:	Identify the project's scope, complexity, timeline, and budget.	Is it a small app development project or a large enterprise software overhaul?
2. Evaluate your team:	Consider team size, skill sets, experience with Agile, and preferred working styles.	Do you have a small, close-knit team or a larger, diverse group?
3. Assess stakeholder involvement:	Determine the level of stakeholder involvement you expect and their need for visibility.	Do stakeholders need frequent updates or prefer minimal interference?

4. Prioritize key factors:	Rank the importance of project complexity, team size, stakeholder involvement, and other relevant aspects.	Is flexibility to adapt to changes most important, or is a predictable delivery schedule crucial?
5. Research & compare methodologies:	Explore different Agile frameworks and their strengths, weaknesses, and ideal project fit.	Use the provided comparison chart and additional resources to learn about Scrum, Kanban, Lean, etc.
6. Consider hybrid approaches:	Don't be afraid to mix and match elements from different methodologies to create a custom fit.	Can you combine Kanban's flow with Scrum's sprint planning for a flexible yet structured approach?
7. Pilot and iterate:	Test-drive your chosen methodology in a small pilot project and adapt based on your team's experience and project feedback.	Start with a short sprint or experiment with Kanban boards before fully committing to a framework.
8. Seek help and feedback:	Consult with Agile coaches,	Discuss your project and

	experienced practitioners, or use online resources for guidance and recommendations.	methodology choice with others to gain valuable insights and refine your approach.

Remember: Choosing the right Agile methodology is an ongoing process. Be open to learning, adapting, and evolving your approach as your project and team dynamics change.

I hope this table provides a clear roadmap for navigating the decision-making process and finding the perfect Agile partner for your project's success!

Remember, Agility is a Journey, Not a Destination

Choosing the right Agile methodology is not just about ticking boxes; it's about **embracing a mindset of continuous learning and adaptation**. As your project evolves, so should your Agile approach. Be open to experimenting, tweaking, and refining your chosen framework to ensure it remains the perfect companion on your Agile journey.

Additional Resources:

- **Scrum Alliance:** https://www.scrumalliance.org/
- **Kanban University:** https://kanban.university/
- **Agile Alliance:** https://www.agilealliance.org/
- **Scaled Agile Framework (SAFe):** https://scaledagileframework.com/

Visual Aids:

- Create a table comparing different Agile methodologies based on factors like complexity, team size, and stakeholder involvement.

- Use a flowchart to illustrate the decision-making process for choosing the right methodology.

By following these tips and delving deeper into the resources provided, you'll be well-equipped to choose the right Agile methodology that propels your project towards success.

Building Effective Agile Teams: The Foundation

Welcome to the cornerstone of your Agile journey! Building a strong, cohesive team is the bedrock upon which successful Agile projects are constructed. In this chapter, we'll delve into the essential ingredients of effective Agile teams, exploring the dynamics, principles, and practices that foster collaboration, trust, and high performance.

The Bedrock Principles of Agile Teams:

- **Individuals and interactions over processes and tools:** Agile values the human element, prioritizing open communication, collaboration, and trust over rigid methodologies. Think of it as building a team of musicians who riff and improvise together, rather than an orchestra bound by a strict conductor's baton.
- **Working software over comprehensive documentation:** Agile embraces iterative development, focusing on delivering functional features and gathering user feedback early and often. Think of it as building a bridge piece by piece, learning from each step and adapting as needed, instead of meticulously drafting blueprints before laying a single brick.
- **Customer collaboration over contract negotiation:** Agile thrives on continuous engagement with stakeholders, ensuring their needs and priorities are woven into the project's fabric. Think of it as co-creating a painting with your client, constantly receiving input and refining your brushstrokes, rather than presenting a finished masterpiece at the end.

- **Responding to change over following a plan:** Agile embraces flexibility and adaptability, readily adjusting to new information, challenges, and market shifts. Think of it as navigating a dynamic river, adjusting course based on the current, rather than clinging to a rigid map.

Building the Agile Team:

Now, let's translate these principles into practical steps for building your Agile team:

1. Assembling the Right Mix:

- **Cross-functionality:** Seek individuals with diverse skills and expertise, fostering a team that can tackle various tasks without relying on external dependencies. Think of it as building a Swiss Army knife team, with members equipped to handle different situations.
- **Communication champions:** Prioritize strong communication skills and active listening. Agile thrives on open dialogue, so team members should be comfortable expressing ideas, asking questions, and providing constructive feedback.
- **Shared ownership and accountability:** Foster a culture where team members feel empowered to take initiative, own their work, and hold themselves and each other accountable. Think of it as building a team of captains, each responsible for steering their section of the ship.

2. Nurturing Trust and Transparency:

- **Psychological safety:** Create an environment where team members feel safe to voice concerns, make mistakes, and learn from them without fear of judgment or retribution. Think of it as building a safe space for experimentation and growth.

- **Open communication channels:** Encourage regular communication through daily stand-up meetings, collaborative tools, and informal interactions. Transparency breeds trust and allows everyone to stay on the same page.
- **Shared goals and vision:** Align the team on a common purpose, ensuring everyone understands their individual roles and how they contribute to the bigger picture. Think of it as painting a mural together, where each brushstroke adds to the collective masterpiece.

3. Fostering Collaboration and Dynamics:

- **Collaborative decision-making:** Encourage collective brainstorming, idea sharing, and consensus-building. Agile thrives on diverse perspectives, so leverage the collective intelligence of your team.
- **Conflict resolution:** Equip your team with healthy conflict resolution skills, enabling them to address disagreements constructively and find mutually beneficial solutions. Think of it as turning conflict into a catalyst for growth and innovation.
- **Team-building activities:** Invest in regular team-building exercises and social interactions. These activities strengthen personal bonds, build trust, and enhance communication outside the project context.

Remember: Building an effective Agile team is an ongoing process. Be prepared to adapt, refine your approach, and learn together as you navigate the Agile journey.

Additional Resources:

- **The Agile Alliance:** https://www.agilealliance.org/
- **HBR Guide to Agile Management:** https://hbr.org/topic/subject/agile-project-management

- **Patrick Lencioni's "Five Dysfunctions of a Team":**
  https://www.amazon.com/five-dysfunctions-team/s?k=five+dysfunctions+of+a+team

By incorporating these principles and practices, you'll be well on your way to building a high-performing Agile team, ready to tackle any project with adaptability, creativity, and success.

Nurturing Agile Teams: Communication and Collaboration

Introduction:

Agile thrives on the dynamic interplay of individuals and interactions. But how do we turn a group of talented individuals into a high-performing Agile team? The secret lies in nurturing communication and collaboration, the lifeblood of Agile success. This chapter dives into the essentials of fostering open communication, effective collaboration, and building a cohesive team culture in your Agile environment.

Breaking Down Communication Barriers:

Effective communication in Agile teams goes beyond information exchange. It's about creating a safe space for open dialogue, active listening, and constructive feedback. Here are some key practices to cultivate:

- **Face-to-Face Communication:** Agile values face-to-face interactions for their immediacy and clarity. Daily stand-up meetings, pair programming sessions, and collaborative workshops encourage real-time exchange of ideas and problem-solving.
- **Transparency and Openness:** Foster a culture of transparency where information flows freely within the team. Share project updates, roadmaps, and challenges openly to ensure everyone is on the same page.
- **Active Listening:** Listen intently to understand, not just to respond. Encourage team members to ask clarifying questions, offer supportive feedback, and be present in conversations.

- **Empathy and Respect:** Build a culture of respect and empathy where everyone feels valued and heard. Different perspectives are encouraged, and disagreements are handled constructively.

Collaboration: More Than Just Teamwork:

Collaboration in Agile goes beyond mere task allocation. It's about working together towards a shared goal, leveraging diverse skills and perspectives. Here are some ways to foster team collaboration:

- **Cross-Functional Teams:** Assemble teams with diverse skillsets to break down silos and encourage cross-functional collaboration. This fosters a holistic understanding of the project and promotes creative problem-solving.
- **Shared Ownership:** Foster a sense of shared ownership where everyone feels invested in the project's success. Encourage team members to contribute ideas, take initiative, and hold each other accountable.
- **Visual Collaboration Tools:** Utilize collaborative tools like whiteboards, Kanban boards, and project management platforms to visualize work progress, share ideas, and track dependencies.
- **Celebrating Wins:** Acknowledge and celebrate team achievements, big or small. This reinforces positive behavior, boosts morale, and strengthens team spirit.

Building a Cohesive Team Culture:

A strong team culture is the cornerstone of effective communication and collaboration. Here are some ways to nurture a positive and productive team environment:

- **Psychological Safety:** Create a space where team members feel safe to take risks, experiment, and voice their

opinions without fear of judgment. Encourage open communication and honest feedback.
- **Trust and Respect:** Build trust through consistent actions, transparency, and open communication. Treat each other with respect, value diverse perspectives, and celebrate individual strengths.
- **Continuous Learning:** Encourage a culture of learning where team members are constantly seeking new knowledge and skills. Share learnings from experiments, failures, and successes to continuously improve as a team.
- **Fun and Social Interaction:** Don't forget the fun! Organize team-building activities, social events, and informal gatherings to foster bonding and strengthen relationships outside of work.

Visual Aids and Examples:

- **Infographic:** Create an infographic illustrating the key elements of effective communication and collaboration in Agile teams.
- **Case Study:** Share a real-world example of how a team overcame communication challenges or leveraged collaboration to achieve success.
- **Scenario-Based Exercises:** Develop interactive exercises where team members can practice communication and collaboration skills in simulated Agile situations.

Additional Resources:

- The Agile Manifesto: https://www.agilealliance.org/agile101/the-agile-manifesto/
- The Art of Agile Development by James Shore and Shane Warden: https://www.amazon.com/Art-Agile-Development-Pragmatic-Software/dp/0596527675

- Radical Candor by Kim Scott:
  https://www.amazon.com/Radical-Candor-Revised-Kick-Ass
  -Humanity/dp/1250235375

Conclusion:

Nurturing communication and collaboration is an ongoing process. By implementing the strategies and tools discussed in this chapter, you can create a thriving Agile team environment where individuals are empowered, ideas flow freely, and success becomes a shared journey. Remember, communication and collaboration are not just tools; they are the very essence of Agile success.

Advanced Team Dynamics in Agile

Introduction:

Having built a strong foundation of communication and collaboration in your Agile team, it's time to delve deeper into the nuanced world of advanced team dynamics. This chapter will explore the subtle factors that take your team from good to great, fostering exceptional levels of trust, psychological safety, and performance.

The Five Dynamics of High-Performing Agile Teams:

1. **Psychological Safety:**
   - **Concept:** Team members feel comfortable taking risks, sharing ideas, and admitting mistakes without fear of judgment or retribution.
   - **Importance:** Psychological safety fosters innovation, learning, and rapid problem-solving.
   - **Building it:** Encourage open communication, celebrate learning from mistakes, and actively listen to diverse perspectives.
2. **Shared Vision and Ownership:**
   - **Concept:** Everyone in the team understands the project's goals and feels ownership over its success.
   - **Importance:** Creates a sense of purpose, alignment, and accountability.
   - **Building it:** Involve the team in goal setting, delegate tasks effectively, and recognize individual and collective achievements.
3. **Continuous Feedback and Improvement:**
   - **Concept:** Regular feedback loops encourage ongoing learning and adaptation, leading to constant improvement.

- **Importance:** Keeps the team agile and responsive to changing needs.
- **Building it:** Implement regular retrospectives, encourage peer feedback, and create a culture of open communication about challenges and opportunities.

4. **Conflict Resolution and Negotiation:**

- **Concept:** Team members can effectively resolve disagreements and negotiate solutions constructively.
- **Importance:** Enables healthy collaboration, prevents misunderstandings, and fosters creative solutions.
- **Building it:** Equip the team with conflict resolution skills, encourage active listening and empathy, and promote win-win solutions.

5. **Self-Organization and Cross-Functionality:**

- **Concept:** Team members are empowered to make decisions, take initiative, and work across functional boundaries.
- **Importance:** Increases efficiency, adaptability, and team autonomy.
- **Building it:** Delegate authority, provide cross-training opportunities, and foster a culture of collaboration and trust.

Examples of Advanced Team Dynamics in Action:

- **Case Study 1:** A software development team implements a rotating "facilitator" role during sprint planning, ensuring everyone has a chance to lead and contribute diverse perspectives.
- **Case Study 2:** A marketing team uses a "challenge and support" approach, where team members feel comfortable questioning each other's ideas while offering constructive feedback and support.
- **Case Study 3:** A design team utilizes a "pair programming" technique, where two designers work together on tasks, fostering knowledge sharing and mutual learning.

Additional Resources:

- **Patrick Lencioni's Five Dysfunctions of a Team:**
  https://www.amazon.com/five-dysfunctions-team/s?k=five+d
  ysfunctions+of+a+team
- **Amy Edmondson's The Fearless Organization:**
  https://www.amazon.com/Fearless-Organization-Psychologi
  cal-Workplace-Innovation-ebook/dp/B07KLT8RKM
- **The Art of Agile Development**:
  https://www.amazon.com/Art-Agile-Development-Pragmatic
  -Software/dp/0596527675

Conclusion:

Mastering advanced team dynamics is a continuous journey, requiring intentional effort and ongoing commitment. By cultivating the five dynamics discussed in this chapter, you can empower your Agile team to reach its full potential, achieving remarkable results and enjoying the journey together.

Remember, building high-performing Agile teams is not about following a rigid formula, but rather fostering a culture of trust, collaboration, and continuous learning. By nurturing these dynamics, you can unlock the magic of Agile and witness your team soar to new heights.

Leading Agile Teams: The Role of a Scrum Master

Agile projects operate differently than traditional ones. They thrive on flexible planning, self-organizing teams, and constant adaptation. But who guides these teams through this dynamic process? Enter the **Scrum Master**, the servant leader who ensures the smooth flow of an Agile project.

Understanding the Scrum Master's Role

Imagine a Scrum Master as an orchestra conductor. They don't play the instruments themselves, but they coordinate the musicians, ensuring harmony and optimal performance. Similarly, Scrum Masters don't directly dictate work; they empower teams to self-organize and collaborate effectively.

Key Responsibilities of a Scrum Master:

- **Facilitating Scrum Events:** Scrum Masters guide the team through essential ceremonies like Sprint Planning, Daily Standups, Sprint Reviews, and Retrospectives. They ensure these events are productive, focused, and adhere to time constraints.
- **Coaching and Mentoring:** Scrum Masters act as coaches, helping team members refine their skills, address challenges, and improve their overall Agile proficiency. This includes fostering open communication, conflict resolution, and continuous learning.
- **Removing Impediments:** Roadblocks are inevitable in any project. Scrum Masters act as shields, protecting the team from distractions and obstacles that impede progress. They identify and remove impediments, allowing the team to stay focused on their goals.

- **Promoting Agile Values and Practices:** Scrum Masters are champions of Agile principles. They educate stakeholders, ensure the team understands and upholds Agile values, and advocate for the adoption of Agile practices within the organization.
- **Monitoring and Adapting:** Scrum Masters constantly monitor the project's progress, analyzing metrics, and identifying areas for improvement. They facilitate adaptations to the Sprint plan based on feedback and learnings, ensuring the team stays on track and delivers value iteratively.

Essential Skills for Scrum Masters:

- **Strong Facilitation Skills:** Scrum Masters need to guide discussions, manage time effectively, and ensure everyone participates actively in Scrum events.
- **Excellent Communication Skills:** Clear and concise communication is crucial for building trust, resolving conflicts, and fostering collaboration within the team.
- **Coaching and Mentoring Expertise:** The ability to guide and support team members, provide constructive feedback, and help them develop their skills is essential.
- **Problem-Solving and Conflict Resolution Skills:** Scrum Masters need to be adept at identifying and resolving roadblocks, addressing team conflicts, and finding creative solutions to challenges.
- **Knowledge of Agile Frameworks:** Understanding the specific Agile framework being used (e.g., Scrum, Kanban) and its intricacies is crucial for effective guidance.

The Scrum Master is not a manager or a team lead. They don't tell people what to do; they empower them to self-organize and take ownership. Their focus is on removing roadblocks, facilitating communication, and ensuring the team has everything they need to succeed.

Examples of Scrum Master Activities:

- **Organizing and facilitating workshops to train the team on Agile principles and practices.**
- **Coaching individual team members on overcoming challenges and improving their skills.**
- **Mediating conflicts within the team and helping them find solutions collaboratively.**
- **Identifying and removing roadblocks, such as resource constraints or dependencies on other teams.**
- **Leading retrospectives to identify areas for improvement and implement changes in the next Sprint.**

Remember, a Scrum Master is not a superhero. They can't solve every problem single-handedly. Their success hinges on building trust and collaboration with the team, creating an environment where everyone feels empowered to contribute and strive for continuous improvement.

Additional Resources:

- Scrum Alliance: https://www.scrumalliance.org/
- Scrum Guides: https://www.scrum.org/resources/scrum-guide
- Agile Alliance: https://www.agilealliance.org/

By understanding the crucial role of the Scrum Master and appreciating their diverse skillset, you'll gain a deeper appreciation for the dynamics of successful Agile teams.

Section 2:
Agile Estimation Techniques

Mastering Agile Estimations: An Introduction

Welcome to the fascinating world of agile estimations! In this chapter, we'll delve into the core concepts of estimating effort and complexity in an agile environment. We'll explore why estimations matter, common techniques used, and how to navigate the challenges of predicting the unpredictable.

Why Estimate in Agile?

Unlike traditional project management, where detailed upfront plans are the norm, agile embraces flexibility and adaptability. However, even in this dynamic setting, estimations play a crucial role. They provide valuable insights into:

- **Prioritizing work:** Knowing the relative effort involved in different tasks helps the team prioritize the product backlog effectively.
- **Sprint planning:** Estimations guide sprint planning by determining how much work can be realistically completed within a sprint timeframe.
- **Managing expectations:** Stakeholders and clients gain a clearer understanding of project timelines and potential roadblocks.

- **Continuous improvement:** Tracking estimations over time allows the team to identify areas for improvement in their planning and execution.

Embracing the "Just Enough" Mindset

Remember, agile estimations are not about achieving perfect accuracy. Instead, the focus is on "just enough" information to make informed decisions. Spending excessive time on precise calculations goes against the agile spirit of rapid iteration and learning.

Popular Agile Estimation Techniques:

Now, let's explore some practical techniques to get you started:

- **Planning Poker:** This interactive method involves team members assigning "story points" to user stories based on relative complexity compared to a reference story. It fosters open discussion and consensus building.

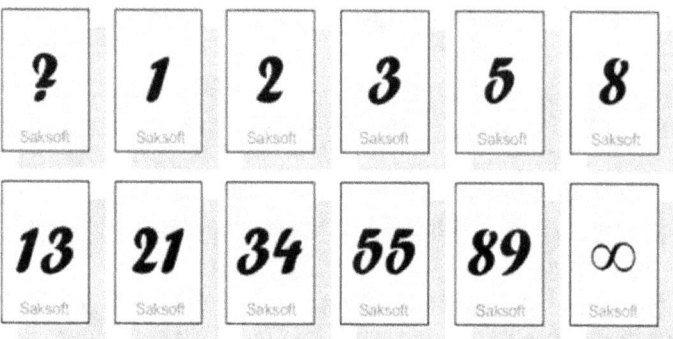

- **T-Shirt Sizing:** This simple method uses pre-defined sizes like XS, S, M, L, XL to categorize user stories based on effort. It's quick and easy to understand, especially for beginners.

Table 1: T-Shirt Sizing Example

Size	Effort Level	Example Tasks
XS	Very small effort	Fix a typo, update documentation
S	Small effort	Implement a simple UI element, write unit tests
M	Medium effort	Develop a new feature with moderate complexity, refactor existing code
L	Large effort	Integrate with a third-party API, build a complex algorithm
XL	Very large effort	Implement a major system component, handle high-risk tasks

- **Fibonacci Sequence:** This sequence (1, 2, 3, 5, 8, 13, …) reflects the diminishing returns of effort as complexity increases. It helps avoid overestimating smaller tasks and underestimating larger ones.

Beyond the Numbers:

Effective estimation goes beyond just assigning points. Remember these key aspects:

- **Focus on relative complexity, not absolute time:** Avoid getting caught up in precise hour calculations. Instead, compare tasks to each other based on effort involved.
- **Embrace collaboration:** Estimations are a team activity, not a solo show. Encourage open discussion, diverse perspectives, and shared understanding.
- **Be flexible and adapt:** As project details evolve, estimations should adapt as well. Don't be afraid to revisit and adjust them during the sprint.
- **Learn from experience:** Track your estimation accuracy over time and identify areas where you can improve. Analyze past projects to understand what worked well and what didn't.

Additional Resources:

- Agile Alliance Estimation Glossary: https://www.agilealliance.org/agile101/agile-glossary/
- Planning Poker at Atlassian: https://marketplace.atlassian.com/apps/1212495/planning-poker
- Agile Estimating and Planning Book by Mike Cohn: https://www.amazon.com/Agile-Estimating-Planning-Mike-Cohn/dp/0131479415

Remember, mastering agile estimations is a continuous journey. Embrace the learning process, experiment with different techniques, and find what works best for your team. By combining these insights with your own agile adventures, you'll be well on your way to conquering the estimation game!

I hope this chapter provides a solid foundation for understanding and applying agile estimations in your projects. Feel free to explore the additional resources for deeper dives into specific techniques and best practices.

Practical Estimation Techniques in Agile

Welcome to the art and science of Agile estimation! In this chapter, we'll dive into the practical toolbox of techniques that help Agile teams predict workload, manage expectations, and navigate the ever-changing landscape of project development. Remember, agile estimations are less about precision and more about fostering collaboration and transparency. Let's explore some **battle-tested methods** to keep your projects on track:

1. T-Shirt Sizing: Embrace the world of fashion! Assign user stories "sizes" like XS, S, M, L, XL (or even 2XL), based on relative complexity compared to other stories. This encourages quick, collaborative estimation without getting bogged down in specific numbers.

Example: During backlog refinement, the team assigns "M" to a user story for adding login functionality, and "S" to a story for fixing a minor UI bug.

2. Planning Poker: Embrace the thrill of the game! Each team member secretly estimates a story using numbered cards (e.g., Fibonacci sequence: 1, 2, 3, 5, 8, etc.). Everyone reveals their cards simultaneously, fostering open discussion and convergence on a final "story point" value.

Example: Team members estimate a story for integrating a payment gateway. One raises a 5 card, another a 3, and a third raises 8. After discussing potential complexities, they settle on a final estimate of 5 story points.

3. Three-Point Estimation: Embrace the power of three! This technique asks for three estimates for each story: optimistic, most likely, and pessimistic. Average the three values to get a more nuanced picture of potential effort involved.

Example: For a story revamping the user interface, the team estimates 2 days (optimistic), 4 days (most likely), and 6 days (pessimistic). The average, 4 days, becomes the story's initial estimate.

4. Affinity Grouping: Embrace the power of sorting! Group similar user stories together based on shared characteristics or complexity. This facilitates relative sizing within each group and promotes consistent estimation across the backlog.

Example: The team groups stories related to user onboarding, e.g., creating an account, verifying email, setting preferences. They then estimate each story within the group relative to others, making sizing more precise.

5. Dot Voting: Embrace the power of democracy! For smaller backlogs, each team member gets a set of "dots" to distribute anonymously among user stories based on perceived complexity. The story with the most dots receives the highest initial estimate.

Example: During backlog refinement, the team uses dot voting to prioritize and estimate a short list of bug fixes. The bug affecting login functionality receives the most dots, indicating its higher complexity and initial estimate.

Remember:

- **No single technique is perfect.** Choose the one that best suits your team's size, project type, and comfort level.
- **Estimates are dynamic.** Regularly revisit and adjust estimations as more information emerges during development.

- **Focus on relative, not absolute accuracy.** Knowing if one story is roughly twice the effort of another is more valuable than pinpointing exact hours.
- **Communication is key.** Discuss assumptions, potential risks, and dependencies during estimation, ensuring everyone's on the same page.

Additional Resources:

- **Agile Alliance Estimation page:** https://hubstaff.com/tasks/agile-estimation-techniques
- **Planning Poker Online Tool:** https://planningpokeronline.com/
- **"Agile Estimating & Planning" by Mike Cohn:** https://www.amazon.com/Agile-Estimating-Planning-Mike-Cohn/dp/0131479415

By mastering these practical estimation techniques, you'll equip your Agile team to navigate the exciting, ever-evolving world of project development with confidence and transparency. Remember, it's not about predicting the future, but about creating a shared understanding of the journey ahead. So, embrace the estimation adventure, and let's get building!

Estimation Pitfalls and How to Avoid Them

Welcome to the treacherous terrain of estimation in agile development! While estimation is crucial for planning and prioritizing work, it's often fraught with challenges. Fear not, intrepid agile warrior, for this chapter equips you with the knowledge to navigate these pitfalls and emerge victorious!

Pitfall 1: The Optimism Bias

Ah, the eternal optimist! We all want to believe in our ability to accomplish great things quickly. Unfortunately, this can lead to underestimating the effort involved in tasks, setting unrealistic expectations, and ultimately, disappointment.

How to dodge it:

- **Use relative estimation:** Compare tasks to similar ones the team has already completed, assigning them a value based on relative complexity (e.g., a "2" if twice as hard as a "1"). This reduces the pressure to predict absolute time and encourages team-based calibration.
- **Plan for buffer time:** Allocate extra time in your sprints to account for unforeseen complexities and unexpected dependencies. Remember, it's better to have buffer you don't need than to scramble under pressure.
- **Embrace the cone of uncertainty:** Acknowledge that early estimates are inherently imprecise. Visualize your estimates as a cone, widening as the task becomes less defined, and narrowing as more information emerges.

Example:

Instead of declaring, "This feature will take 2 days," say, "This feature feels like a '3' based on its complexity compared to the '2' we spent on the login page. Let's revisit the estimate as we learn more."

Pitfall 2: Anchoring Bias

Our minds tend to latch onto the first piece of information we receive, anchoring our subsequent estimates to it. This can be problematic when the initial input is inaccurate or irrelevant.

How to avoid it:

- **Estimate in silence:** Before discussing a task, have each team member independently jot down their estimates. This eliminates the influence of others' opinions and encourages individual critical thinking.
- **Use Fibonacci sequences:** Instead of linear numbers, use a Fibonacci sequence (1, 2, 3, 5, 8, 13) for estimation. This encourages larger jumps in complexity between values, reducing the tendency to over-refine estimates.
- **Challenge assumptions:** Don't blindly accept the first explanation of a task. Question its scope, complexity, and potential dependencies to break free from anchoring bias.

Example:

Instead of immediately agreeing with someone who says, "This is a quick 1-point task," ask clarifying questions like, "Are there any API integrations involved?" or "Does it require front-end changes?"

Pitfall 3: The Planning Fallacy

We tend to underestimate the time it takes to complete tasks, often neglecting unforeseen obstacles and distractions. This can lead to overloaded sprints and burnout.

How to conquer it:

- **Break down tasks into smaller chunks:** Divide complex tasks into smaller, more manageable units, making it easier to estimate and track progress.
- **Factor in distractions:** Don't assume you'll have uninterrupted development time. Account for meetings, emergencies, and other interruptions that can disrupt your flow.
- **Use historical data:** Analyze your team's past performance on similar tasks to identify patterns and adjust your estimates accordingly.

Example:

Instead of estimating the entire "User Management System" feature as a single task, break it down into smaller units like "User registration form," "User authentication," and "User profile editing," then estimate each individually.

Pitfall 4: Ignoring Dependencies

Tasks rarely exist in isolation. Dependencies between tasks can significantly impact their timelines. Neglecting these dependencies can lead to delays and frustration.

How to stay ahead:

- **Map dependencies explicitly:** Clearly identify dependencies between tasks and visualize them on a Kanban board or dependency map. This allows the team to see the big picture and plan accordingly.
- **Prioritize dependent tasks:** Ensure tasks that others rely on are completed first to avoid bottlenecks and keep the flow of work moving smoothly.
- **Communicate dependencies effectively:** Keep stakeholders informed about dependencies and potential

delays caused by them. Transparency fosters understanding and helps manage expectations.

Example:

Before starting on the "Payment Integration" task, ensure the "User Account" functionality is complete, as it's a prerequisite. Communicate this dependency to both development and product teams to avoid roadblocks.

Bonus Tip: Adopt a culture of continuous improvement. Regularly analyze your estimation accuracy and identify areas for improvement. Don't be afraid to adapt your methods and tools as you learn and gain experience.

Remember, accurate estimation is a continuous journey, not a one-time destination. By recognizing and avoiding these pitfalls, you can equip your agile team to conquer estimation challenges and deliver value with greater predictability and confidence.

Additional Resources:

- **Agile Estimating and Planning by Mike Cohn:** https://www.amazon.com/Agile-Estimating-Planning-Mike-Cohn/dp/0131479415
- **The Art of Agile Development by James Shore and Shane Warden:** https://www.amazon.com/Art-Agile-Development-Pragmatic-Software/dp/0596527675
- **Agile Project Management with Scrum by Ken Schwaber and Jeff Sutherland:** https://www.scrum.org/resources/scrum-guide

Section 3:
Agile Planning Strategies

Why Traditional Plans Fail in Agile Projects

Introduction:

In the world of project management, the term "plan" conjures images of Gantt charts, timelines, and detailed milestones. These rigid structures are the backbone of traditional project management methodologies, but in the dynamic world of Agile, they often become an obstacle to success. This chapter delves into the fundamental reasons why traditional plans fail in Agile projects, helping you avoid these pitfalls and embrace the flexibility and adaptability that Agile champions.

The Mismatch between Assumptions and Reality:

Traditional project management thrives on upfront planning, assuming a predictable environment with clearly defined requirements. However, Agile projects operate in a dynamic landscape where change is the norm, not the exception. This fundamental mismatch leads to several problems:

- **Inaccurate Estimates:** Traditional methods rely on fixed estimates made at the project's outset, often based on incomplete information. As requirements evolve and unforeseen challenges arise, these estimates quickly become obsolete, causing frustration and budget overruns.

- **Scope Creep:** Rigid plans struggle to accommodate changes in scope, leading to feature creep or scope reduction, both of which negatively impact project outcomes. Agile, on the other hand, embraces iterative development, allowing for scope adjustments throughout the project lifecycle.
- **Limited Visibility:** Traditional plans provide a limited view of progress, often masking underlying issues until it's too late. Agile's focus on transparency and frequent feedback loops ensures early detection and course correction, preventing small problems from snowballing into major roadblocks.

The Silos of Roles and Responsibilities:

Traditional project management often creates distinct roles with segregated responsibilities. This siloed approach can hinder collaboration and communication, essential aspects of successful Agile teams.

- **Lack of Ownership:** In a traditional setting, developers work on assigned tasks without a deep understanding of the overall project goals. This can lead to disengagement and a lack of ownership, which negatively affects project quality. Agile teams, on the other hand, foster cross-functional collaboration and empower individuals to take ownership of their work, leading to increased motivation and better results.
- **Inefficient Communication:** Traditional communication channels can be slow and cumbersome, hindering the rapid feedback loops crucial for Agile success. Agile teams utilize daily stand-up meetings, backlog refinement sessions, and visual communication tools to ensure everyone is on the same page and problems are addressed promptly.

The Fear of Change:

Traditional project managers are often trained to minimize risk and stick to the plan. However, Agile projects thrive on embracing change and adapting to new information. This can lead to resistance to:

- **Evolving Requirements:** Clinging to outdated requirements can lead to a product that doesn't meet customer needs. Agile teams embrace iterative development, allowing for continuous refinement of requirements based on user feedback and changing priorities.
- **Experimentation and Learning:** Traditional approaches view experimentation as a risk to be minimized. Agile, however, encourages experimentation as a way to learn and improve the product. This iterative process leads to faster innovation and better outcomes.

Examples of Traditional Plans Gone Wrong:

Consider a traditional software development project where a detailed timeline and budget are set based on initial requirements. Months into the project, the client discovers a critical feature is missing. Adding this feature requires significant changes to the plan, leading to budget overruns, schedule delays, and frustration.

In contrast, an Agile team working on the same project would have identified the missing feature during a sprint review and quickly adjusted their backlog to accommodate it. The iterative nature of Agile would allow them to adapt their plan without derailing the project, ultimately delivering a product that meets the client's evolving needs.

Table comparing traditional and Agile planning approaches

Feature	Traditional Planning	Agile Planning
Planning Duration	Upfront, detailed plan for the entire project	Iterative, ongoing planning within each sprint
Assumptions	Predictable environment, fixed requirements	Dynamic environment, evolving requirements
Estimates	Fixed estimates made at the outset	Continuous refinement of estimates based on progress and feedback
Focus	Meeting deadlines and adhering to the plan	Delivering value and adapting to change
Communication	Siloed communication channels	Frequent feedback loops and open communication

Approach to Change	Minimizing and resisting change	Embracing and adapting to change

Key Takeaways:

Traditional plans fail in Agile projects because they are based on assumptions that don't hold true in the dynamic world of Agile. To succeed, Agile teams need to embrace flexibility, collaboration, and continuous learning. By understanding the pitfalls of traditional planning and adopting Agile principles, you can navigate the ever-changing landscape of your project and deliver successful outcomes.

Further Resources:

- **The Agile Manifesto:**
  https://www.agilealliance.org/agile101/the-agile-manifesto/
- **Scrum Guide:** https://scrumguides.org/
- **Kanban University:** https://kanban.university/
- **"Agile Project Management" by Andrew Stellman and Jen Greene:**
  https://www.amazon.com/agile-project-management-Books/s?k=agile+project+management&rh=n%3A283155

Remember, the key to success in Agile is not about avoiding plans altogether, but about adopting a flexible and adaptable approach that allows you to learn, adjust, and deliver value iteratively.

Crafting an Agile Plan: Flexibility and Adaptability

Welcome to the dynamic world of Agile planning! In this chapter, we'll dive into the heart of crafting an Agile plan, emphasizing flexibility and adaptability as the cornerstones of success.

Traditional Waterfall vs. Agile Planning: A Paradigm Shift

Before we delve into Agile planning, let's refresh our memory with the limitations of traditional waterfall planning. Imagine a cascading waterfall, where each stage (requirements, design, development, testing, deployment) must be completed in sequence before moving to the next. This rigid approach often leads to:

- **Delays and missed deadlines:** Unforeseen changes late in the project cycle can cause ripple effects, pushing back timelines and straining resources.
- **Reduced customer feedback:** Limited opportunities for customer input throughout the project restrict adaptation to changing needs and market demands.
- **Low team morale:** Working in silos with lengthy stages can be demotivating for teams, hindering collaboration and innovation.

Agile planning flips the script! Instead of a waterfall, picture a nimble river, flowing and adapting to the terrain. Agile plans embrace:

- **Iterative development:** Work is broken down into smaller, time-boxed iterations (sprints), allowing for continuous feedback, learning, and adjustments.

- **Prioritization and flexibility:** Features are prioritized based on value and business needs, with the freedom to re-evaluate and adapt priorities as the project progresses.
- **Continuous collaboration:** Agile teams work closely together throughout the project, fostering communication, shared ownership, and rapid problem-solving.

The Agile Planning Toolbox: Principles and Frameworks

Now, let's equip ourselves with the tools for crafting a flexible and adaptable Agile plan.

1. Vision and Goals:

- **Start with a clear vision:** Define your project's ultimate aim and desired outcomes. This provides a North Star for decision-making throughout the project.
- **Set SMART goals:** Specific, Measurable, Achievable, Relevant, and Time-bound goals ensure focus and alignment with the overall vision.

2. Prioritization and Backlog Management:

- **Prioritize ruthlessly:** Use techniques like MoSCoW (Must-have, Should-have, Could-have, Won't-have) to prioritize features based on value and impact.
- **Maintain a dynamic backlog:** The backlog is a prioritized list of all project tasks and features. Keep it flexible and updated regularly to reflect evolving priorities and new information.

3. Iterative Planning and Sprints:

- **Break down work into sprints:** Each sprint typically lasts 1-4 weeks, focusing on delivering a specific set of features and learning from them.

- **Plan effectively with sprint planning meetings:** These collaborative sessions involve the entire team defining sprint goals, tasks, and estimates.

4. Embrace Change and Continuous Improvement:

- **Expect and welcome change:** Don't be afraid to adapt your plan based on new information, feedback, and market shifts.
- **Conduct regular retrospectives:** These post-sprint reflection sessions help identify successes, challenges, and areas for improvement, feeding continuous learning and optimization.

Examples of Agile Planning Frameworks:

- **Scrum:** A popular framework with sprint cycles, roles like Scrum Master and Product Owner, and rituals like daily standups and sprint reviews.
- **Kanban:** A visual framework that uses boards and cards to represent work in progress, highlighting bottlenecks and promoting continuous flow.
- **Lean Startup:** Emphasizes rapid prototyping, validation, and learning through iterative cycles of build-measure-learn.

Remember, the key to successful Agile planning is not creating a rigid roadmap, but establishing a flexible framework that guides your journey while allowing you to adapt to the ever-changing landscape of your project.

Additional Resources:

- Agile Alliance: https://www.agilealliance.org/
- Scrum Guides: https://www.scrum.org/resources/scrum-guide
- Kanban University: https://kanban.university/
- Lean Startup: https://theleanstartup.com/

Managing Scope and Expectations in Agile

One of the most common challenges faced by Agile teams is managing the scope and expectations of their projects. Unlike traditional project management methodologies where scope is fixed, Agile is designed to be flexible and open to change. This can often lead to misaligned expectations between the team and stakeholders, resulting in missed targets, delays, and missed opportunities.

In this chapter, we'll examine how Agile teams can effectively manage the scope and expectations of their projects, while still maintaining the core principles of Agile.

Understanding Scope and Expectations

Before we dive into the strategies for managing scope and expectations in Agile, let's first define what these terms mean in the context of project management.

Scope refers to the work that needs to be done to deliver the project's objectives. In an Agile project, scope is fluid and open to change. As the team learns more about the requirements and objectives of the project, the scope may change, be added to, or be removed.

Expectations, on the other hand, are what stakeholders anticipate the project will deliver. These can be related to budget, timelines, functionality, or other aspects of the project.

To effectively manage scope and expectations in Agile, teams need to have a deep understanding of both.

Agile Techniques for Managing Scope and Expectations

1. Prioritization: Agile teams must have a process for prioritizing work based on its impact on the project's objectives. This can help manage both scope and expectations. Stakeholders can see the progress of the project and understand that the team is working on the most critical items first. This can lead to a better alignment of expectations and an increased understanding of the project's goals.

2. Understanding Acceptance Criteria: Agile teams need to understand the acceptance criteria for project tasks. This will ensure that they are working towards specific deliverables and can help manage stakeholder expectations. By setting clear acceptance criteria, the team can minimize potential scope creep and ensure that the project stays on track.

3. Managing Change Requests: Agile teams must have a process for managing change requests. This process should include how to approve change requests, how to document the changes, and how to communicate the changes to stakeholders. By managing change requests effectively, the team can ensure that stakeholders are aware of changes in scope and can adjust their expectations accordingly.

4. Continuous Communication: Agile teams must communicate regularly with stakeholders throughout the project. This can help manage expectations, provide visibility into the project's progress, and ensure that stakeholder feedback is incorporated into the project.

5. Setting Realistic Deadlines: Agile projects often use time-boxed iterations, which can make it challenging to set specific deadlines. However, it's essential to have some form of a timeline to manage expectations. By setting realistic deadlines, teams can ensure that stakeholders understand the project's progress and can make informed decisions.

Conclusion

Managing scope and expectations in Agile is critical to the success of the project. By prioritizing work, understanding acceptance criteria, managing change requests, communicating regularly, and setting realistic deadlines, teams can manage stakeholder expectations effectively. This can lead to a better alignment of expectations and ultimately result in delivering a successful project.

Creating Your First Agile Project Plan

Introduction

Creating your first Agile project plan is a pivotal step in managing Agile projects. This chapter aims to guide you through the process of developing an Agile project plan, emphasizing flexibility, iterative development, and collaboration.

The Essence of Agile Planning

Agile planning differs fundamentally from traditional project planning. It's not about setting a rigid project path but about creating a flexible roadmap that can adapt to changes and feedback.

Key Principles:

1. **Iterative Planning**: Breaking down the project into smaller, manageable iterations.
2. **Adaptive Schedules**: Adjusting timelines based on team progress and stakeholder feedback.
3. **Collaborative Effort**: Involving the team and stakeholders in the planning process.
4. **Value-Driven Delivery**: Prioritizing tasks that deliver the most value to the customer first.

Step-by-Step Guide to Creating Your Agile Project Plan

Step 1: Define Project Vision and Goals

- **Vision Statement**: Define what success looks like for the project.

- **SMART Goals**: Specific, Measurable, Achievable, Relevant, Time-bound goals.

Step 2: Build the Product Backlog

- List all features, functions, and tasks required for the project.
- Use user stories to describe features from the customer's perspective.
- Prioritize the backlog based on value delivery and dependencies.

Step 3: Estimate Tasks

- Employ techniques like Planning Poker or T-shirt sizing for estimations.
- Estimations should be in terms of effort and complexity, not just time.

Step 4: Define Iterations/Sprints

- Divide the project into short iterations (usually 2-4 weeks).
- For each iteration, select a set of tasks from the product backlog.
- Ensure each iteration delivers a potentially shippable product increment.

Step 5: Plan Iteration Details

- Assign specific tasks to team members.
- Discuss and remove any roadblocks.
- Set iteration goals and key deliverables.

Step 6: Review and Adapt the Plan

- Regularly review the plan with your team.
- Adjust future iterations based on feedback and learnings.
- Maintain an up-to-date project roadmap.

Communication and Collaboration

Effective communication is vital in Agile planning. Regular stand-ups, iteration planning meetings, and retrospective sessions ensure everyone is aligned and the plan stays relevant.

Tools and Resources

- **Agile Project Management Tools**: Jira, Trello, Asana
- **Estimation Tools**: Planning Poker, Fibonacci Sequence cards
- **Collaboration Tools**: Slack, Microsoft Teams

Examples and Case Studies

- **Case Study 1**: A software development team using Agile planning to adapt to changing customer needs.
- **Case Study 2**: An e-commerce project where Agile planning helped manage a tight schedule and budget constraints.

Conclusion

Creating your first Agile project plan is about embracing change, focusing on delivering value, and fostering team collaboration. Remember, the plan is a living document that guides your project journey rather than dictates it.

Additional Resources:

1. "Agile Estimating and Planning" by Mike Cohn - for deep insights into Agile planning and estimation techniques.
2. "Succeeding with Agile" by Mike Cohn - for broader concepts of implementing Agile methodologies.

3. Scrum Guides (scrumguides.org) - for understanding the fundamentals of Scrum, a key Agile framework.

Utilizing Burndown Charts for Effective Tracking

Introduction

In Agile project management, tracking progress is crucial for the timely and successful completion of projects. Burndown charts are pivotal tools in Agile that offer a visual representation of work left versus time. This chapter delves into the nuances of using burndown charts effectively to track Agile project progress.

What is a Burndown Chart?

A burndown chart is a graphical representation that shows the amount of work remaining in a project over time. It is a simple yet powerful tool used in Agile methodologies, particularly in Scrum, to gauge the progress of a project iteration, known as a Sprint.

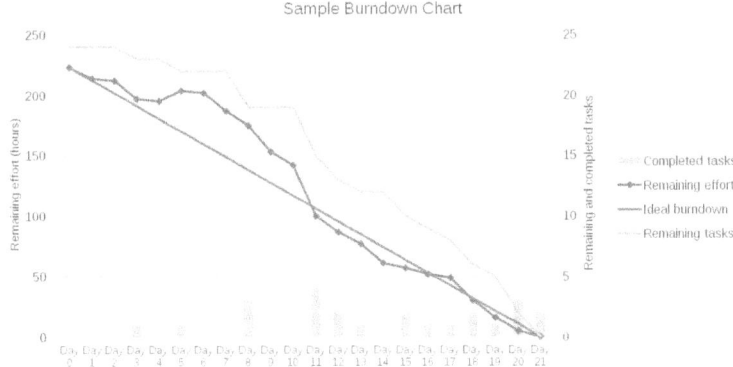

Sample Burndown Chart

Elements of a Burndown Chart

- **X-Axis (Time):** Represents the timeline of the Sprint, usually in days.

- **Y-Axis (Work Remaining):** Indicates the amount of work remaining, often measured in story points or hours.
- **Ideal Burndown Line:** Shows the ideal work rate required to complete the work on time.
- **Actual Burndown Line:** Reflects the actual progress of the team.

Creating a Burndown Chart

1. **Estimate the Total Workload:** Initially, the team estimates the total workload, often in story points or hours.
2. **Daily Updates:** Each day, the team updates the chart by subtracting the work completed from the total workload.
3. **Plotting Progress:** The remaining workload is plotted against the time, forming the Actual Burndown Line.

Using Burndown Charts for Effective Tracking

1. **Identifying Deviations:** Comparing the actual progress line with the ideal line helps identify deviations from the plan.
2. **Early Warning System:** Sudden changes in the slope of the Actual Burndown Line can act as an early warning for potential issues.
3. **Adapting Plans:** Based on the chart, teams can adapt their strategies to meet the Sprint goals.

Best Practices

- **Update Regularly:** Updating the chart daily ensures accuracy in tracking.
- **Avoid Scope Creep:** The chart helps in visualizing scope creep when the line trends upwards.
- **Team Involvement:** Encourage team involvement in updating and analyzing the chart for shared understanding and responsibility.

Common Challenges and Solutions

- **Estimation Errors:** Over- or under-estimation of tasks can skew the chart. Regular retrospective meetings can help in refining estimation skills.
- **Unexpected Workload Increases:** Sudden increases in workload should be discussed and addressed in stand-up meetings.

Case Study: Effective Use of Burndown Charts

Consider a case where a software development team is working on a new feature. The burndown chart initially shows a lag in the team's progress. By analyzing the chart, the team realizes that the lag is due to underestimation of a complex task. The team decides to break down the task into smaller, manageable parts and revises the sprint backlog accordingly. This action reflects in the burndown chart as the Actual Line starts to align more closely with the Ideal Line, helping the team to stay on track.

Conclusion

Burndown charts are indispensable tools in Agile project management, providing clear, visual cues about the project's progress and health. By effectively utilizing burndown charts, Agile teams can enhance their ability to track progress, identify potential issues early, and adapt their plans to ensure successful Sprint completion.

Additional Resources

For further reading and examples on utilizing burndown charts in Agile projects, consider the following resources:

1. **"Agile Estimating and Planning" by Mike Cohn:** Offers a comprehensive understanding of planning in Agile, including the use of burndown charts.

2. **Scrum.org Resources:** Provides a range of articles, case studies, and webinars on Agile practices including burndown charts.
3. **Agile Alliance Website:** Contains a wealth of resources including experience-based reports on the use of burndown charts in various Agile environments.

Examples for Illustration

- **Example Burndown Chart Templates:** Online platforms like Atlassian and Trello offer templates that can be used to create and understand burndown charts.
- **Interactive Burndown Chart Simulators:** Some online tools allow users to simulate burndown charts with varying parameters, helping to understand how different factors affect the project's progress.

Case Studies: Burndown Charts in Action

Introduction

In this chapter, we delve into real-world applications of burndown charts in Agile project management. Burndown charts, a visual tool for tracking work completion against time, are vital in Agile planning and execution. We will explore case studies to illustrate their practical use and effectiveness in various scenarios.

Case Study 1: Software Development Project

Background: A mid-sized tech company embarks on a new software development project. The project is scheduled to last three months, with a team of ten members.

Application of Burndown Chart:

- **Week 1:** The initial chart shows a steep decline as the team completes a significant portion of tasks.
- **Mid-Project Challenge:** A critical bug is discovered, causing the chart to plateau. The team recalibrates the work plan.
- **Result:** The burndown chart reflects this adjustment with a less steep slope, but progress continues. The project completes two days behind schedule but remains within acceptable limits.

Lesson Learned: The burndown chart provided early warning signs of the delay, allowing for timely intervention.

Case Study 2: Marketing Campaign

Background: An advertising agency employs Agile for a six-week marketing campaign.

Application of Burndown Chart:

- **Variable Task Completion:** Unlike software development, creative tasks had less predictable completion times.
- **Adaptation:** The team adjusted the chart weekly to reflect the true pace of work.
- **Result:** The campaign was a success, with the burndown chart serving as a guide for balancing creative effort and deadlines.

Lesson Learned: Burndown charts must be adaptable to the nature of the work and team dynamics.

Case Study 3: Agile in Education

Background: A university adopts Agile methodologies for curriculum development over a semester.

Application of Burndown Chart:

- **Long-term Project:** The burndown chart was set for a four-month period, with milestones.
- **Unexpected Delays:** Faculty strikes and administrative changes caused significant deviations.
- **Result:** Regular updates to the chart and Agile ceremonies helped realign the team's focus, leading to successful curriculum development despite setbacks.

Lesson Learned: In long-term projects, burndown charts assist in maintaining focus and adjusting to changes.

Using Burndown Charts Effectively

1. **Regular Updates:** Update charts regularly to reflect true progress.

2. **Flexibility:** Be prepared to adjust the trajectory in response to unforeseen events.
3. **Communication Tool:** Use the chart as a basis for daily standups and planning meetings.
4. **Beyond Software Development:** Understand that burndown charts are useful in various industries, not just IT.

Conclusion

These case studies demonstrate the versatility and effectiveness of burndown charts in Agile project management across different sectors and project types. By providing a visual representation of progress and challenges, burndown charts serve as a powerful tool for guiding Agile teams toward successful project completion.

Additional Resources

1. "Agile Estimating and Planning" by Mike Cohn - Offers an in-depth understanding of Agile planning and estimation, including the use of burndown charts.
2. "Scrum: The Art of Doing Twice the Work in Half the Time" by Jeff Sutherland - Provides insights into Scrum methodology and its tools.
3. Online Course: "Agile with Atlassian Jira" on Coursera - A practical course for implementing Agile techniques using Jira software, including burndown charts.

The Planning Workshop: Interactive Learning in Agile Planning

Introduction

Welcome to the chapter on "The Planning Workshop: Interactive Learning in Agile Planning," a critical component in the "Agile Planning Strategies" section of this book. Agile Planning is an iterative and collaborative process, and the Planning Workshop is its cornerstone. This chapter will delve into the concept, execution, and benefits of an effective Planning Workshop in Agile Projects.

1. Understanding the Agile Planning Workshop

1.1 Definition and Objectives

An Agile Planning Workshop is a collaborative event involving team members, stakeholders, and sometimes customers. It aims to define and prioritize tasks, set short-term goals, and establish a clear roadmap for the project.

1.2 Key Components

- **Product Backlog**: A prioritized list of project requirements and features.
- **Sprint Goals**: Specific objectives for the upcoming sprint.
- **Task Breakdown**: Splitting features into manageable tasks.

Example: Imagine a software development team working on a new application. In the Planning Workshop, they would break down the feature 'User Registration' into tasks like 'Design UI for Registration', 'Set up Database for User Data', etc.

2. Conducting a Successful Planning Workshop

2.1 Preparing for the Workshop

- **Participant Selection**: Ensure a mix of skills and perspectives.
- **Agenda Setting**: Define clear objectives and outcomes for the workshop.
- **Material Preparation**: Equip the space with necessary tools like whiteboards, markers, and sticky notes.

2.2 Workshop Activities

- **User Story Mapping**: Creating a visual representation of user journeys.
- **Estimation Techniques**: Using methods like Planning Poker to estimate effort.
- **Prioritization**: Utilizing MoSCoW or other techniques to prioritize tasks.

Resource: For an in-depth understanding of User Story Mapping, Jeff Patton's book "User Story Mapping" is highly recommended.

2.3 Facilitation Tips

- **Encourage Participation**: Ensure every team member's voice is heard.
- **Manage Time**: Keep discussions focused and within allotted time slots.
- **Document Outcomes**: Record decisions and action items.

3. Interactive Learning in Agile Planning

3.1 The Role of Interactive Learning

Interactive learning in Agile Planning involves hands-on activities, real-time feedback, and collaborative problem-solving, leading to a deeper understanding of project goals and enhanced team cohesion.

3.2 Techniques for Interactive Learning

- **Role-Playing**: Simulating real-world scenarios.
- **Group Discussions**: Brainstorming and collective problem-solving.
- **Games and Simulations**: Using Agile games to understand concepts.

Example: A role-playing session where team members act out different stakeholder roles to understand their needs and expectations.

---

4. Benefits and Challenges

4.1 Benefits

- **Improved Understanding**: Enhanced grasp of project scope and requirements.
- **Team Collaboration**: Strengthened team dynamics and communication.
- **Adaptability**: Ability to respond to changes and feedback effectively.

4.2 Challenges and Mitigation

- **Time Constraints**: Prioritize agenda items and keep discussions focused.
- **Diverse Opinions**: Foster an environment of respect and constructive feedback.

- **Documentation Overload**: Use digital tools for efficient recording.

Conclusion

The Planning Workshop in Agile is more than a meeting; it's an interactive, collaborative, and dynamic process that sets the stage for a successful Agile journey. By engaging in interactive learning and effectively conducting these workshops, teams can ensure that they are well-prepared to tackle the challenges of Agile projects with a clear, shared vision and strategy.

Additional Resources

1. *"Agile Estimating and Planning"* by Mike Cohn - For deeper insights into Agile planning techniques.
2. Online Course: *"Agile Planning for Software Products"* - Offers practical examples and exercises.
3. Blog: *"Effective Facilitation in Agile Workshops"* - Tips and tricks for successful workshop facilitation.

Section 4:
Executing Agile Iterations

The Anatomy of an Agile Iteration

Introduction

Agile iterations, often referred to as sprints in Scrum, are the heartbeat of any Agile project. They provide a structured yet flexible framework for teams to continuously deliver value to their customers. This chapter will dissect an Agile iteration, illustrating its various components and their interplay to facilitate effective project execution.

1. Iteration Planning

1.1 Overview

Iteration planning is the inception point of any Agile iteration. It involves the whole team and aims to determine what work will be done during the iteration.

1.2 Key Activities

- **Backlog Refinement**: Prioritizing and estimating backlog items.
- **Setting Iteration Goals**: Defining clear, achievable objectives for the iteration.

- **Capacity Planning**: Assessing the team's available bandwidth.
- **Selecting Stories**: Choosing backlog items that align with iteration goals and capacity.

1.3 Example

Imagine a software development team working on a new feature for a mobile app. During iteration planning, they select user stories related to enhancing the user interface, based on their priority and the team's capacity.

2. Daily Stand-ups

2.1 Overview

Daily stand-ups are brief, time-boxed meetings where team members synchronize their work and report on obstacles.

2.2 Key Activities

- **Progress Update**: Each member summarizes what they did yesterday, what they plan to do today, and any impediments.
- **Identifying Blockers**: Team collaboratively identifies solutions to remove any impediments.

2.3 Example

A team member reports they are blocked by a dependency on an external API. The team discusses potential workarounds or alternative tasks until the blocker is resolved.

3. Iteration Execution

3.1 Overview

This is where the team works on the tasks defined in the iteration plan, adhering to Agile principles and practices.

3.2 Key Activities

- **Task Breakdown and Execution**: Breaking down stories into tasks and working on them.
- **Continuous Collaboration**: Regularly interacting with team members and stakeholders.
- **Adaptive Planning**: Adjusting tasks and priorities as new information emerges.

3.3 Example

A developer finds an unanticipated technical challenge. The team collaborates to reassess the approach, ensuring they still meet the iteration goals.

4. Continuous Integration and Testing

4.1 Overview

In Agile, work is continuously integrated and tested to ensure quality and functionality.

4.2 Key Activities

- **Automated Testing**: Implementing automated unit, integration, and acceptance tests.
- **Regular Code Integration**: Merging code changes frequently to avoid integration issues.
- **Continuous Feedback Loop**: Gathering feedback from automated tests and stakeholders.

4.3 Example

After implementing a new feature, the automated test suite runs, revealing a few bugs. The team addresses these issues promptly, ensuring the product's quality.

5. Iteration Review and Demonstration

5.1 Overview

The iteration review is a meeting where the team showcases what they've accomplished during the iteration.

5.2 Key Activities

- **Demonstrating Completed Work**: Presenting completed stories to stakeholders.
- **Gathering Feedback**: Receiving input from stakeholders on the work done.

5.3 Example

The team demonstrates the new user interface features to stakeholders, who suggest minor adjustments for the next iteration.

6. Iteration Retrospective

6.1 Overview

The retrospective is a reflective meeting held at the end of each iteration to discuss what went well and what could be improved.

6.2 Key Activities

- **Reflecting on Processes**: Analyzing both successes and failures in processes.
- **Actionable Improvements**: Identifying specific areas for improvement in the next iteration.
- **Team Bonding**: Strengthening team dynamics through open and constructive dialogue.

6.3 Example

The team realizes that communication gaps have delayed some tasks. They decide to update their communication protocol for better clarity in the future.

The Anatomy of an Agile Iteration Table

Component	Description
Planning	The team defines the goals for the iteration, prioritizes and breaks down work into user stories, and estimates the effort involved.
Development	The team works on completing the user stories, following Agile practices like daily stand-up meetings, pair programming, and

	continuous integration and delivery.
Testing & Quality Assurance	The team tests the work completed throughout the iteration to ensure it meets quality standards.
Delivery & Deployment	The completed work is delivered to stakeholders, often in the form of a working demo or potentially shippable product.
Feedback & Retrospective	The team reflects on the iteration, identifies what worked well and what could be improved, and uses this feedback to plan for the next iteration.

Here are some additional details about each component:

- **Planning:** This is typically done in a sprint planning meeting, where the team collaborates to define the iteration's goals, scope, and backlog.
- **Development:** Agile teams often use Scrum or Kanban boards to visualize their work and track progress. Daily

stand-up meetings help ensure everyone is on track and identify any roadblocks.

- **Testing & Quality Assurance:** Testing is integrated throughout the development process, not just at the end. Automated testing tools can help to catch bugs early.
- **Delivery & Deployment:** The goal is to deliver working software early and often, so that stakeholders can provide feedback and the team can learn from their mistakes.
- **Feedback & Retrospective:** The retrospective is a crucial opportunity for the team to learn and improve. It's important to be honest and open about what worked well and what didn't, and to make concrete plans for improvement in the next iteration.

Conclusion

Understanding the anatomy of an Agile iteration is crucial for successfully executing Agile projects. Each component plays a vital role in maintaining the iterative, collaborative, and adaptive nature of Agile methodologies. By mastering these elements, teams can efficiently navigate the complexities of Agile projects, delivering high-quality products that meet customer needs and expectations.

Additional Resources

For further reading and in-depth understanding, consider the following resources:

1. **"Agile Estimating and Planning" by Mike Cohn:** Offers comprehensive insights into planning and estimating in Agile projects.

2. **"Succeeding with Agile" by Mike Cohn**: Provides a detailed exploration of implementing Agile methodologies in organizations.

3. **"Continuous Delivery: Reliable Software Releases through Build, Test, and Deployment Automation" by Jez Humble and David Farley**: Essential reading for understanding continuous integration and deployment practices in Agile.

Implementing Agile Iterations: A Practical Guide

Welcome to the engine room of your Agile journey! In this chapter, we'll delve into the nitty-gritty of **executing Agile iterations**, the repetitive yet dynamic cycles that breathe life into your project. We'll equip you with the practical steps and tools needed to navigate each iteration with confidence, delivering value and adapting to change with grace.

Setting the Stage:

Before diving into the specifics, let's establish the foundation. Each iteration is a timeboxed sprint, typically lasting 1-4 weeks. This timeframe provides a focused burst of activity, allowing you to deliver working software and gather valuable feedback early and often. Remember, Agile is all about **embracing change and iteration**, so be prepared to adapt as you learn.

Essential Steps for Each Iteration:

1. Iteration Planning:

- **Prioritize the Backlog:** Collaborate with your team to prioritize the backlog of user stories or tasks. Focus on high-value items that deliver tangible results within the iteration timeframe.
- **Break Down the Work:** Divide the prioritized backlog items into smaller, manageable tasks. This ensures clarity, facilitates estimation, and prevents teams from feeling overwhelmed.
- **Set the Iteration Goal:** Define a clear and measurable goal for the iteration. This could be delivering a specific feature, reaching a certain milestone, or achieving a performance improvement.

2. Daily Stand-up Meetings:

- **Short and Focused:** Daily stand-up meetings should be quick (15 minutes max) and focused on three key questions: what did I do yesterday, what will I do today, and are there any roadblocks?
- **Promote Transparency:** These meetings foster open communication, identify potential issues early, and allow for quick course correction.
- **Keep it Standing:** Encourage standing to maintain energy and focus.

3. Development and Testing:

- **Embrace Continuous Integration and Delivery (CI/CD):** Automate as much of the development and testing process as possible to ensure frequent integration and rapid feedback.
- **Prioritize Quality:** Implement effective testing practices throughout the iteration, including unit testing, integration testing, and user acceptance testing.
- **Collaborate and Share:** Encourage knowledge sharing and cross-functional collaboration within the team.

4. Iteration Review and Retrospective:

- **Review the Results:** Reflect on the iteration's successes and challenges. Did you achieve the goals? What went well? What could have been improved?
- **Identify Learnings:** Use the retrospective to identify key learnings and actionable insights for the next iteration.
- **Celebrate Achievements:** Don't forget to acknowledge and celebrate the team's hard work and accomplishments.

Tooling and Techniques:

- **Agile Project Management Tools:** Utilize tools like Jira, Trello, or Asana to visualize the workflow, track progress, and manage the backlog.
- **Kanban Boards:** Implement a Kanban board to visualize the flow of work, limit work in progress, and identify bottlenecks.
- **Retrospective Techniques:** Use techniques like the Five Whys, Mad Sad Glad, or Starfish to facilitate productive retrospectives and action planning.

Remember:

- **Embrace Flexibility:** Agile is about adapting to change, so be prepared to adjust your plans as you learn and receive feedback.
- **Focus on Value:** Prioritize delivering value to the customer with each iteration.
- **Communicate Effectively:** Open and transparent communication is key to success in any Agile project.

Conclusion:

Implementing Agile iterations effectively requires a balance of planning, execution, and adaptation. By following the practical steps outlined in this chapter and utilizing the right tools and techniques, you can navigate each iteration with confidence, delivering value and building a successful Agile project.

Bonus Tip: Encourage a culture of learning and experimentation within your team. This will foster continuous improvement and ensure that your Agile iterations are always evolving and growing.

Overcoming Challenges in Agile Iterations: Navigating the Rapids of Change

Agile iterations, with their emphasis on rapid development and continuous improvement, are a powerful engine for innovation. But these fast-paced cycles also present unique hurdles. Navigating these challenges requires a blend of resilience, adaptability, and a proactive approach. This chapter delves into the most common obstacles encountered during agile iterations, equipping you with the tools and strategies to overcome them and emerge stronger.

1. Scope Creep and Unrealistic Expectations

Agile embraces flexibility, but uncontrolled scope creep can quickly derail even the most well-planned iteration. Stakeholders may introduce new requirements mid-sprint, or the team might underestimate the complexity of existing tasks. To combat this:

- **Prioritization is Key:** Establish a clear product vision and prioritize features based on their value and impact. Utilize techniques like MoSCoW (Must-Have, Should-Have, Could-Have, Won't-Have) to manage expectations.
- **Embrace the MVP:** Deliver a Minimum Viable Product (MVP) with core functionalities in the first iteration. This allows for early feedback and prioritizes value delivery over feature bloat.
- **Continuous Communication:** Maintain open communication channels with stakeholders. Regularly review progress, update expectations, and negotiate changes within the sprint's constraints.

2. Time Management and Missed Deadlines

Agile sprints operate on fixed timelines, but unforeseen challenges can lead to missed deadlines. This can erode trust and create a sense of panic. Remember:

- **Realistic Estimation:** Conduct thorough task estimation and utilize historical data to improve accuracy. Employ techniques like relative sizing or planning poker to involve the whole team.
- **Buffer Time:** Allocate buffer time within the sprint to absorb unforeseen delays and unexpected tasks. This ensures core functionalities are delivered even when minor setbacks occur.
- **Adapt and Prioritize:** If deadlines become impossible to meet, prioritize tasks based on impact and value. Be prepared to defer less critical features to the next iteration.

3. Lack of Communication and Collaboration

Agile thrives on open communication and collaboration. However, siloed teams, unclear roles, and poor communication can lead to misunderstandings and inefficiencies. To foster a collaborative environment:

- **Daily Stand-up Meetings:** Utilize daily stand-up meetings to share progress, identify roadblocks, and ensure alignment. Keep them concise and focused on actionable items.
- **Visual Management Tools:** Employ visual tools like Kanban boards or task lists to track progress and dependencies transparently. This fosters accountability and keeps everyone on the same page.
- **Retrospectives and Continuous Improvement:** Regularly conduct retrospectives to identify areas for improvement, celebrate successes, and address communication gaps. Encourage open dialogue and feedback loops.

4. Quality Concerns and Defects

Agile prioritizes rapid delivery, but quality should never be compromised. Defects and bugs discovered in production can damage reputation and erode user trust. To ensure high-quality software:

- **Shift-Left Testing:** Integrate testing throughout the development cycle, not just at the end. Utilize automated testing frameworks and continuous integration/continuous delivery (CI/CD) pipelines to catch issues early.
- **Pair Programming and Code Reviews:** Encourage pair programming and code reviews to identify potential bugs and improve code quality. This shared responsibility fosters knowledge sharing and collective ownership of quality.
- **Refactoring and Continuous Improvement:** Don't be afraid to refactor code during iterations. This improves maintainability, reduces technical debt, and prevents quality degradation over time.

5. Resistance to Change and Cultural Shift

Agile methodologies can be a significant change from traditional waterfall approaches. This can lead to resistance and discomfort among team members accustomed to set routines. To overcome this:

- **Invest in Training and Education:** Provide training and resources that educate team members about agile principles and benefits. This helps build understanding and reduce fear of the unknown.
- **Lead by Example:** Agile leadership requires open communication, transparency, and a willingness to adapt. Demonstrate the value of agile practices through your own behavior.

- **Celebrate Successes:** Recognize and celebrate successful iterations and achievements. This reinforces the positive aspects of agile and encourages continued adoption.

Remember, overcoming challenges in agile iterations is an ongoing process. By embracing a proactive approach, fostering open communication, and continuously learning and adapting, your team can navigate the rapids of change and deliver high-quality software faster and more efficiently.

Section 5:
Agile Communication and Feedback

Essential Communication Practices in Agile Iterations

Introduction

Agile project management emphasizes effective communication as a cornerstone of successful project execution. This chapter delves into essential communication practices within Agile iterations, focusing on fostering a transparent, collaborative, and adaptive environment. These practices are integral to the "Agile Communication and Feedback" section of our book, emphasizing their role in the broader Agile framework.

Understanding Agile Communication

Agile Communication: Unlike traditional models, Agile communication is continuous, collaborative, and iterative. It's not just about sharing information but about creating a shared understanding among the team.

Importance: Effective communication in Agile ensures that team members are aligned with the project goals, understand changes quickly, and can adapt to new information effectively.

Key Communication Practices in Agile Iterations

1. **Daily Stand-Up Meetings**
   - **Purpose**: To synchronize the team's activities and plan for the next 24 hours.
   - **Structure**: Short, time-boxed meetings where each member answers three questions: What did I do yesterday? What will I do today? Are there any impediments in my way?
   - **Example**: A developer might report they completed a feature implementation yesterday, plan to work on bug fixes today, and mention a blocker due to an external dependency.
2. **Iteration Planning**
   - **Purpose**: To plan the work for the iteration.
   - **Structure**: The team selects items from the backlog to work on during the iteration.
   - **Example**: The team might decide to prioritize customer-reported bugs in the next iteration.
3. **Backlog Refinement**
   - **Purpose**: To review and revise the product backlog.
   - **Structure**: The team discusses the backlog items, estimates efforts, and clarifies requirements.
   - **Example**: Clarifying user stories or splitting larger items into smaller, manageable tasks.
4. **Retrospectives**
   - **Purpose**: To reflect on the past iteration and identify areas for improvement.
   - **Structure**: Discussing what went well, what didn't, and what can be improved.
   - **Example**: The team might realize that communication with a stakeholder needs improvement.
5. **Demonstrations/Reviews**

- ○ **Purpose**: To showcase the work completed during the iteration.
- ○ **Structure**: The team presents completed work to stakeholders for feedback.
- ○ **Example**: Demonstrating a new feature to the product owner and other stakeholders.

Enhancing Communication Effectiveness

1. **Visual Tools**: Use of Kanban boards, burndown charts, and other visual aids to provide a clear picture of progress and issues.
2. **Active Listening**: Encouraging team members to listen actively to understand and respond effectively.
3. **Feedback Loops**: Establishing continuous feedback mechanisms with stakeholders.
4. **Conflict Resolution**: Addressing and resolving conflicts promptly to maintain team harmony.
5. **Tooling**: Utilizing tools like JIRA, Slack, or Trello to streamline communication.

Overcoming Communication Challenges

1. **Remote Teams**: Use of video conferencing and collaboration tools to bridge the physical gap.
2. **Cultural Differences**: Fostering an inclusive environment that respects diverse backgrounds and opinions.
3. **Information Overload**: Prioritizing communication to avoid overwhelming team members.

Conclusion

Effective communication in Agile iterations is not just about talking more; it's about talking better. Through the practices outlined in this chapter, Agile teams can ensure that they are not only productive but also collaborative and adaptive to change.

Additional Resources

- **Agile Communication**: "Crucial Conversations: Tools for Talking When Stakes Are High" by Kerry Patterson et al.
- **Tooling for Agile Teams**: A comparison of Agile tools on [Capterra] (https://www.capterra.com/agile-project-management-software/).
- **Effective Retrospectives**: "Agile Retrospectives: Making Good Teams Great" by Esther Derby and Diana Larsen.

Rethinking the Daily Standup: Best Practices

Introduction

The daily standup, a staple of Agile methodologies, is more than just a routine meeting. It's a pivotal tool for ensuring team alignment, identifying roadblocks, and fostering a culture of collaboration. However, to extract the maximum value from these meetings, we need to rethink and optimize our approach.

The Essence of Daily Standups in Agile

Daily standups, typically a part of the Scrum framework, are short, time-boxed meetings (usually 15 minutes) where team members quickly discuss their progress since the last meeting, plans for the day, and any impediments they're facing.

Key Objectives:

1. Synchronize team efforts.
2. Identify and address roadblocks.
3. Foster a collaborative environment.

Best Practices for Effective Standups

1. **Timing is Key**: Schedule the standup at a time that works for everyone. Consistency is important to establish a routine.
2. **Stand Up, Stay Brief**: Encourage standing up to keep the meeting short and on point. This helps maintain energy and focus.
3. **Focus on Progress, Not Status**: Shift the conversation from status updates to progress discussion. Emphasize what has been accomplished and what will be done next.

4. **Address Roadblocks Immediately**: When impediments are mentioned, note them but discuss resolutions post-standup to avoid prolonging the meeting.
5. **Rotate the Facilitator**: This encourages engagement and gives everyone a chance to lead.
6. **Use a Physical or Digital Board**: Visual aids, like Kanban or Scrum boards, help keep the team focused and discussions relevant.
7. **Keep it Public and Open**: Invite stakeholders but maintain the focus on the team. Stakeholders should observe and not participate unless necessary.
8. **End with a Positive Note**: A quick round of recognition or a positive note can boost team morale.

Rethinking the Standup Format

- **Walking the Board**: Instead of the traditional three questions, discuss items based on their current status on the board.
- **Themed Standups**: Occasionally focus on specific areas like risks, learning, or continuous improvement.
- **Asynchronous Standups**: For distributed teams, consider using tools where team members can post updates at a fixed time.

Incorporating Feedback Loops

Regularly seek feedback on the standup's effectiveness and be open to experimenting with formats, times, and structures to find what works best for the team.

Common Pitfalls to Avoid

- Dominance by a few individuals.
- Turning standups into problem-solving sessions.
- Irrelevant or too detailed discussions.
- Inconsistent attendance or engagement.

Additional Resources

1. *Scrum Guide* – Offers a detailed overview of Scrum practices.
2. *Agile Retrospectives: Making Good Teams Great* – Provides insights into continuous improvement, which can be applied to enhancing standup meetings.
3. [Kanban and Scrum Board Software] – Tools like Jira, Trello, or Asana for digital board management.

Conclusion

Rethinking the daily standup involves shifting focus from mere status updates to fostering collaboration, addressing impediments quickly, and continuously adapting the format to suit the team's evolving needs. By applying these best practices, your standup meetings can transform into a powerful tool for agile success.

The Feedback Loop Workshop: Continuous Improvement in Agile

Introduction

Welcome to Chapter 24 of "Agile Crash Course," where we delve into the critical concept of continuous improvement in Agile through the Feedback Loop Workshop. This chapter, situated within the "Understanding Agile" section of the book, aims to provide an in-depth understanding of how feedback loops function in Agile environments and how to effectively implement them in your projects.

The Essence of Feedback in Agile

Agile methodologies thrive on iterative development and continuous feedback. The feedback loop is an integral component that drives improvement in products, processes, and teams. It involves regularly gathering feedback from all stakeholders and promptly acting on it to enhance future iterations.

Setting Up the Feedback Loop Workshop

1. **Objective Setting**: Define clear goals for the workshop. This includes identifying specific areas for feedback, such as product features, process efficiency, or team dynamics.
2. **Participant Selection**: Involve diverse participants, including team members, stakeholders, and customers, to gather a broad range of perspectives.
3. **Creating a Safe Environment**: Ensure an atmosphere where participants feel comfortable sharing honest and constructive feedback.

Conducting the Workshop

1. **Facilitation Techniques**: Use skilled facilitators to guide the workshop, ensuring it stays focused and productive.
2. **Feedback Collection Methods**: Implement various methods like surveys, open discussions, and retrospective meetings to collect feedback.
3. **Recording and Organizing Feedback**: Document the feedback systematically for future reference and action.

Analyzing Feedback

1. **Identifying Themes**: Look for common patterns or themes in the feedback.
2. **Prioritization**: Prioritize feedback based on its impact on project goals and resource availability.
3. **Developing Action Plans**: Create specific, actionable plans to address the feedback.

Implementing Changes

1. **Integrating Feedback into Iterations**: Incorporate the actionable feedback into upcoming iterations.
2. **Communicating Changes**: Clearly communicate the changes being made in response to feedback to all stakeholders.

Measuring Improvement

1. **Setting Metrics**: Establish clear metrics to measure the effectiveness of changes made.
2. **Reviewing Progress**: Regularly review these metrics to assess if the changes are yielding the desired results.

Challenges and Best Practices

1. **Overcoming Resistance**: Address resistance to change by emphasizing the value of feedback in improving outcomes.
2. **Ensuring Continuity**: Establish a culture where feedback is continuously sought, analyzed, and acted upon.
3. **Best Practices**: Share best practices and learnings from successful feedback implementations.

Conclusion

The Feedback Loop Workshop is a powerful tool in the Agile toolkit. It not only helps in enhancing the product but also fosters a culture of openness, collaboration, and continuous improvement. By effectively implementing these workshops, Agile teams can ensure that they are always moving towards better processes, products, and team dynamics.

Additional Resources

To further enrich your understanding of the Feedback Loop Workshop in Agile, the following resources are recommended:

1. **Books**:
    o "Agile Retrospectives: Making Good Teams Great" by Esther Derby and Diana Larsen
    o "The Lean Startup" by Eric Ries (specifically the chapters on feedback loops and iterative development)
2. **Online Courses**:
    o "Agile Retrospectives: The Complete Guide" (Available on platforms like Coursera or Udemy)
    o "Effective Feedback in Agile Teams" (Offered by various Agile coaching institutes)
3. **Websites**:
    o [Agile Alliance] (https://www.agilealliance.org/): Resources and articles on Agile methodologies.

- ○ [Scrum.org] (https://www.scrum.org/): Insights and articles on implementing Scrum and feedback loops.
4. **Forums and Communities**:
     - ○ Join Agile and Scrum forums on platforms like LinkedIn or Reddit for real-world advice and discussions.

These resources will offer comprehensive insights and practical knowledge to deepen your understanding and application of feedback loops in Agile environments.

Section 6:
Agile in Software Development

The Role of Unit Testing in Agile

Introduction

In the Agile world, where rapid and iterative development is key, Unit Testing emerges as a cornerstone practice. This chapter delves into how unit testing fits into the Agile methodology, particularly in software development, ensuring robust, reliable, and adaptive software solutions.

The Essence of Unit Testing in Agile

Unit testing, in its simplest form, involves testing the smallest parts of an application independently, usually methods or functions. In Agile, this practice aligns seamlessly with iterative development. Each iteration or sprint often introduces new functionalities or refines existing ones. Unit tests ensure that these incremental changes do not break the software, providing a safety net for continuous improvement.

Benefits of Unit Testing in Agile

1. **Early Bug Detection:** Bugs are identified and fixed during the development phase, long before they become expensive and complex to resolve.

2. **Refactoring Confidence:** Developers can refactor code with assurance, as tests will reveal if changes alter the desired behavior.
3. **Documentation:** Well-written unit tests act as documentation, showing how the code is supposed to work.
4. **Simplified Integration:** Unit testing reduces integration issues, as each part is verified to work correctly before integration.
5. **Quality Assurance:** Continuous testing inherently elevates the overall quality of the software.

Writing Effective Unit Tests

Effective unit tests are:

- **Independent:** Tests should not rely on external factors or other tests.
- **Repeatable:** They should produce the same results every time, regardless of the environment.
- **Small:** Covering small parts of functionality makes them easier to write and maintain.
- **Readable:** Tests should be clear and understandable, serving as documentation.
- **Fast:** They need to run quickly to not slow down the development process.

Integrating Unit Testing in Agile Workflows

1. **Test-Driven Development (TDD):** Begin with writing a failing test before writing the code to pass the test. It ensures that testing is at the forefront.
2. **Incorporate in Definitions of Done:** A user story or feature is not 'done' until all its unit tests pass.
3. **Continuous Integration (CI):** Integrate unit tests into CI pipelines. Tests are run automatically whenever changes are made, ensuring immediate feedback.

Common Challenges and Solutions

- **Legacy Code Without Tests:** Start by writing tests for any new code and gradually add tests to legacy code during refactoring.
- **Time Constraints:** Prioritize unit testing for critical parts of the application. Over time, the time saved by reduced bug fixing outweighs the initial investment.
- **Resistance to Change:** Educate teams on the benefits and provide training. Make testing a part of the team culture.

Tools and Frameworks

There are various tools and frameworks available for unit testing, such as JUnit for Java, NUnit for .NET, and Jest for JavaScript. These tools provide a structure for writing and running tests, along with useful features like mock objects and assertions.

Conclusion

Unit testing is an integral part of Agile software development, ensuring that each part of the application is reliable and functions as intended. By integrating unit testing into daily workflows, teams can enjoy faster development cycles, higher code quality, and reduced bug fixing time.

Additional Resources

For more in-depth knowledge, readers are encouraged to explore:

1. "Test Driven Development: By Example" by Kent Beck
2. "Clean Code: A Handbook of Agile Software Craftsmanship" by Robert C. Martin
3. Online course: "Unit Testing for Agile Software Development" on platforms like Coursera or Udemy
4. Agile Testing Community forums and meetups for real-world advice and networking.

Refactoring Techniques for Agile Teams

Introduction to Refactoring in Agile

Refactoring is a systematic process of improving existing software code without altering its external behavior. In the Agile context, it's crucial for maintaining the adaptability and quality of code. This chapter delves into various refactoring techniques suitable for Agile teams, emphasizing iterative improvements and collaborative efforts.

Understanding the Need for Refactoring in Agile

1. **Maintaining Code Quality**: Over time, code can become complex and difficult to understand. Refactoring helps in keeping the codebase clean and manageable.
2. **Enhancing Code Adaptability**: Agile projects evolve rapidly, and the code must adapt to changing requirements. Refactoring ensures that the code remains flexible.
3. **Facilitating Continuous Integration**: Regular refactoring is essential in a continuous integration environment to prevent integration issues.

Core Refactoring Principles for Agile Teams

- **Incremental Changes**: Small, frequent refactorings are preferable over large, infrequent ones.
- **Test-Driven Approach**: Ensure each refactoring step is validated by tests to maintain functionality.

- **Collective Code Ownership**: Encourage team members to contribute to refactoring efforts.
- **Continuous Learning**: Use refactoring as an opportunity for team members to learn from each other's techniques.

Effective Refactoring Techniques

1. **Simplifying Conditional Statements**: Break down complex conditions into simpler, more readable forms.
2. **Method Extraction**: Replace repeated code blocks with a single method.
3. **Renaming for Clarity**: Rename variables, methods, and classes to reflect their purpose more clearly.
4. **Removing Dead Code**: Eliminate unused code to reduce complexity and confusion.
5. **Reducing Coupling**: Modify code to reduce dependencies between classes.
6. **Refactoring for Reusability**: Modify code to make it reusable in different contexts.
7. **Object-Oriented Refactoring**: Use object-oriented principles to enhance code modularity and readability.

Best Practices for Agile Team Refactoring

- **Refactor During Downtime**: Identify opportunities for refactoring during periods of less intense development.
- **Peer Review and Pair Programming**: Engage in peer review and pair programming for collective decision-making and knowledge sharing.
- **Document Refactoring Decisions**: Keep a record of why certain refactoring decisions were made for future reference.

- **Balance Refactoring with New Feature Development**: Ensure that refactoring efforts do not overshadow the progress of new features.

Tools and Resources

- **Refactoring Tools**: Automated tools like ReSharper, JRefactory, or RubyMine can assist in identifying and performing refactorings.
- **Code Analysis Tools**: Tools like SonarQube help in identifying complex code that might benefit from refactoring.

Conclusion

Refactoring is an essential practice for Agile software development teams, ensuring that the codebase remains clean, understandable, and adaptable to change. By adhering to the principles and techniques outlined in this chapter, Agile teams can enhance their efficiency and maintain high-quality software products.

Additional Resources

1. **"Refactoring: Improving the Design of Existing Code" by Martin Fowler**: A comprehensive guide on refactoring techniques.
2. **Online Course: "Refactoring for Software Engineers"**: Offers practical, real-world examples of refactoring in Agile environments.
3. **Agile Alliance Resources on Refactoring**: Provides a collection of articles and case studies on Agile refactoring.

Advanced Refactoring Strategies

Introduction

In previous chapters, we explored the fundamental principles of Agile and its application in software development, with a particular focus on the crucial role of refactoring. As we delve into advanced refactoring strategies, it's essential to remember that these techniques are not just about altering code; they're about improving the design, readability, and maintainability of software while keeping it fully functional.

Understanding Advanced Refactoring

Advanced refactoring goes beyond simple code modification. It involves:

1. **Architectural Refactoring**: Restructuring the software's architecture without changing its external behavior. This might involve changing the software's modular structure, improving data flow, or enhancing the separation of concerns.
2. **Design Pattern Refactoring**: Implementing design patterns to solve specific design problems or to enhance code readability and reusability.
3. **Legacy Code Refactoring**: Dealing with code that lacks proper tests. It involves identifying critical sections of the code, writing tests, and then safely refactoring.

Strategies for Effective Advanced Refactoring

1. **Test-Driven Development (TDD)**: Before refactoring, write tests for the existing functionality. This ensures that the refactored code still meets the requirements.

2. **Small, Incremental Changes**: Make small changes and test frequently. Large changes increase the risk of introducing errors.
3. **Continuous Integration (CI)**: Regularly integrate changes into the main branch to identify integration issues early.
4. **Pair Programming**: Two heads are often better than one. Pair programming allows for real-time code review and idea sharing.
5. **Tool Support**: Utilize refactoring tools for automated code analysis and refactoring tasks. However, don't rely solely on these tools; manual review is crucial for understanding the broader impact of changes.

Advanced Refactoring Techniques

1. **Composing Methods**: Break down large methods into smaller, more manageable pieces.
2. **Simplifying Conditional Expressions**: Reducing complexity in if-else structures.
3. **Refactoring for Reusability**: Making code more modular and reusable.
4. **Dealing with Data Clumps**: Grouping related data together into a class or structure.
5. **Refactoring Inheritance Hierarchies**: Simplifying complex inheritance structures.

Case Studies and Examples

Example 1: Refactoring a monolithic application into a microservices architecture.

Example 2: Implementing the Strategy pattern to replace conditional logic in a pricing module.

Best Practices and Pitfalls

- **Do Not Refactor Everything at Once**: Prioritize what needs refactoring based on the project's goals.
- **Maintain Codebase Integrity**: Ensure that the application remains in a working state after each refactoring session.
- **Avoid Refactoring During Critical Phases**: Avoid major refactoring close to a release or during other critical project phases.

Conclusion

Advanced refactoring is a powerful tool in maintaining and enhancing the quality of software. It requires a strategic approach, where understanding the code, collaborating with the team, and focusing on testing are key.

Additional Resources

For further reading and resources, consider the following:

1. *Refactoring: Improving the Design of Existing Code* by Martin Fowler
2. Online courses on advanced refactoring techniques (e.g., Coursera, Udemy)
3. Software refactoring tools and plugins for integrated development environments (IDEs)

By mastering advanced refactoring strategies, Agile teams can significantly improve the longevity and performance of their software products.

Introduction to Test-Driven Development (TDD)

Welcome to Chapter 28 of "Agile Crash Course" where we delve into one of the most transformative practices in agile software development: Test-Driven Development (TDD). As part of our journey through Agile in Software Development, this chapter aims to demystify TDD and equip you with the knowledge to implement it effectively in your agile projects.

What is Test-Driven Development?

Test-Driven Development is a software development approach where tests are written before the actual code. The core idea is simple: create a test for a specific piece of functionality, watch it fail (since the functionality doesn't exist yet), write the minimal code needed to pass the test, and then refactor the code while ensuring it still passes the test.

The TDD Cycle: Red, Green, Refactor

1. **Red:** Write a test that defines a function or improvements of a function, which should fail initially because the feature doesn't exist.
2. **Green:** Write the minimal code necessary to pass the test. The focus here is on functionality, not perfection.
3. **Refactor:** Clean up the code while keeping it functional. This step focuses on code quality, removing duplication, and improving performance.

Benefits of TDD in Agile Projects

- **Enhanced Code Quality:** TDD leads to a more thoughtful design and cleaner code, as it requires developers to consider requirements before writing code.

- **Reduces Bugs:** Early detection of defects makes it easier to manage and rectify issues, leading to more robust software.
- **Facilitates Change:** With a robust suite of tests, developers can refactor and upgrade systems with confidence, knowing that their changes do not break existing functionality.
- **Improves Documentation:** Since tests describe how the software should behave, they act as a form of documentation for the system.

Implementing TDD in Agile Teams

1. **Mindset Shift:** Embrace the TDD philosophy. It's not just about testing; it's about developing a system through tests.
2. **Incremental Development:** Start small. Create tests for small units of functionality and expand gradually.
3. **Collaboration:** Encourage collaboration within the team. Pair programming can be particularly effective in a TDD environment.
4. **Continuous Integration:** Integrate TDD with Continuous Integration practices for constant feedback and early detection of issues.

Challenges and Solutions

- **Initial Learning Curve:** TDD can be counterintuitive at first. Solution: Provide training and allow time for developers to adapt.
- **Increased Initial Effort:** Writing tests before code requires more upfront effort. Solution: Highlight long-term benefits like reduced maintenance costs.
- **Resistance to Change:** Changing established workflows can meet resistance. Solution: Demonstrate the benefits through small pilot projects.

TDD in Action: Case Studies

- **Case Study 1:** A startup adopting TDD saw a 40% reduction in bug reports after six months.
- **Case Study 2:** An enterprise team transitioned to TDD and experienced improved collaboration and quicker feature releases.

Conclusion

As we conclude this chapter on Test-Driven Development, remember that TDD is not just a testing approach; it's a comprehensive design philosophy. It encourages thoughtful design, results in cleaner and more maintainable code, and aligns perfectly with the iterative, incremental nature of agile development.

Additional Resources

For further reading and a deeper understanding of Test-Driven Development, consider the following resources:

1. "Test-Driven Development: By Example" by Kent Beck.
2. "Growing Object-Oriented Software, Guided by Tests" by Steve Freeman and Nat Pryce.

Setting Up Continuous Integration: The Basics

Introduction

Welcome to Chapter 30, "Setting Up Continuous Integration: The Basics." This chapter is a pivotal part of the "Agile in Software Development" section of our book. Continuous Integration (CI) is a cornerstone in modern agile software development, offering numerous benefits like early bug detection, enhanced code quality, and a smoother development process.

Understanding Continuous Integration

Continuous Integration is a development practice where developers frequently integrate their code into a shared repository, ideally several times a day. Each integration is then verified by an automated build and automated tests. The primary goals of CI are to identify issues and conflicts early, reduce integration problems, and ensure the quality of the software being developed.

Key Principles:

1. **Automate the Build:** Every code commit should trigger an automated build and test sequence.
2. **Maintain a Single Source Repository:** All code should be committed to a shared repository.
3. **Build Every Commit:** This ensures immediate feedback on any integration issues.
4. **Keep the Build Fast:** A fast build encourages frequent commits.
5. **Test in a Clone of the Production Environment:** This ensures consistency between development and production.

Setting Up a Continuous Integration Environment

Step 1: Choose a CI Server

- *Examples:* Jenkins, Travis CI, CircleCI, and GitLab CI.
- *Considerations:* Integration with existing tools, scalability, and community support.

Step 2: Establish a Source Code Repository

- *Options:* GitHub, GitLab, Bitbucket.
- *Best Practices:* Use branches for features, bug fixes, and releases.

Step 3: Configure the CI Server

- Integrate the CI server with your repository.
- Define build triggers (e.g., on commit, periodically).
- Set up notifications for build outcomes.

Step 4: Create Build Scripts

- Define steps for compiling, testing, and packaging your software.
- Use a build tool like Maven, Gradle, or Ant.

Step 5: Write Automated Tests

- Essential for detecting problems early.
- Include unit tests, integration tests, and code quality checks.

Step 6: Monitor and Optimize

- Regularly review build reports and test outcomes.
- Optimize build times and address flaky tests.

Best Practices for Continuous Integration

1. **Commit Regularly:** Small, frequent commits reduce integration complexity.
2. **Fix Broken Builds Immediately:** This maintains the integrity of the codebase.
3. **Write Clean, Testable Code:** Ensures that automated tests are effective.
4. **Use Feature Flags:** Allows merging of incomplete features without affecting functionality.
5. **Continuously Refine and Improve:** CI is an ongoing process, not a set-and-forget system.

Common Challenges and Solutions

- **Integration Conflicts:** Encourage more frequent commits and enhance communication among developers.
- **Slow Builds:** Optimize build scripts, and consider parallelizing tests.
- **Flaky Tests:** Focus on improving test reliability and repeatability.

Conclusion

Setting up Continuous Integration is a fundamental step towards a more streamlined and efficient development process in agile environments. By following the steps and best practices outlined in this chapter, you'll lay a solid foundation for a robust CI process in your project.

For additional resources and deeper understanding, consider exploring the following:

- [Continuous Integration: Improving Software Quality and Reducing Risk] (https://www.amazon.com/Continuous-Integration-Improving -Software-Reducing/dp/0321336380) by Paul M. Duvall, Steve Matyas, and Andrew Glover.

- [Jenkins: The Definitive Guide]
  (https://www.oreilly.com/library/view/jenkins-the-definitive/97
  81449305352/) by John Ferguson Smart.
- Online tutorials and courses on platforms like Pluralsight,
  Coursera, or Udemy focused on specific CI tools like
  Jenkins, Travis CI, or CircleCI.

Remember, the journey to effective continuous integration is
iterative and evolves with your project. Keep learning,
experimenting, and improving.

Optimizing Continuous Integration: Advanced Strategies

Introduction

In the rapidly evolving landscape of software development, Continuous Integration (CI) stands as a cornerstone of Agile methodologies. CI is more than just a practice; it's a philosophy that enables teams to integrate changes frequently, detect errors early, and deliver quality software efficiently. While the basics of CI set the stage for smoother development cycles, optimizing CI requires a deeper understanding and strategic approach. This chapter delves into advanced strategies to optimize your CI processes, ensuring they align with Agile principles and contribute significantly to the success of your software projects.

Understanding the Essence of CI in Agile

Before diving into advanced strategies, it's crucial to reaffirm the role of CI in Agile software development. CI is not just about automating the build and testing processes; it's about fostering a culture where small, frequent changes are integrated, tested, and delivered quickly. This approach minimizes integration issues, enhances collaboration, and aligns perfectly with Agile's iterative nature.

Key Components of CI:

1. **Automated Builds:** Ensuring that code changes are automatically built and errors are detected early.
2. **Automated Testing:** Running tests automatically to ensure new changes don't break the application.

3. **Immediate Feedback:** Providing real-time feedback to developers about the state of their code.

Advanced Strategies for Optimizing CI

1. Pipeline as Code

- **Concept:** Treat your CI/CD pipeline configuration as code, allowing it to be version-controlled and reviewed like any other code.
- **Benefits:** Improves traceability, enhances collaboration, and allows for easy rollback of pipeline changes.

2. Parallel Execution and Distributed Builds

- **Techniques:** Splitting tests into smaller chunks and running them in parallel; distributing builds across multiple machines.
- **Advantages:** Reduces build times, accelerates feedback loops, and improves resource utilization.

3. Advanced Branching Strategies

- **Approaches:** Implementing strategies like feature toggles and trunk-based development to manage code branches effectively in a CI environment.
- **Outcome:** Facilitates continuous integration of code changes, minimizing merge conflicts and integration issues.

4. Environment Consistency

- **Implementation:** Using containerization and infrastructure as code (IaC) to maintain consistent environments across development, testing, and production.
- **Result:** Reduces the "it works on my machine" syndrome and ensures reliable, reproducible builds.

5. Flaky Test Management

- **Methods:** Identifying, isolating, and addressing flaky tests (tests that produce inconsistent results).
- **Impact:** Increases confidence in the CI process and reduces time wasted on false positives.

6. Security Integration

- **Integration:** Embedding security checks and scans within the CI pipeline.
- **Significance:** Ensures that security is a continuous concern, leading to more secure software.

7. Performance Testing in CI

- **Implementation:** Incorporating performance testing in the CI pipeline to detect performance regressions early.
- **Advantages:** Ensures that performance considerations are addressed continuously, not just at the end of the development cycle.

8. Monitoring and Analytics

- **Application:** Implementing monitoring tools to track the health and performance of the CI pipeline.
- **Benefit:** Provides insights for continuous improvement and helps in proactively addressing pipeline bottlenecks.

Optimizing Continuous Integration is not just about speeding up the build process; it's about creating a more robust, efficient, and collaborative development environment. By implementing these advanced strategies, teams can fully embrace the Agile philosophy, delivering high-quality software at a faster pace.

Additional Resources for Further Reading

1. **"Continuous Delivery: Reliable Software Releases through Build, Test, and Deployment Automation"** by Jez Humble and David Farley - A comprehensive guide on CI/CD practices.
2. **"Accelerate: The Science of Lean Software and DevOps: Building and Scaling High Performing Technology Organizations"** by Nicole Forsgren, Jez Humble, and Gene Kim - Insights into high-performing IT organizations.
3. **Docker and Kubernetes Documentation** - For understanding containerization and orchestration in CI/CD.
4. **"Effective DevOps: Building a Culture of Collaboration, Affinity, and Tooling at Scale"** by Jennifer Davis and Katherine Daniels - Insights into building a DevOps culture that aligns with CI/CD principles.